OFFICIAL SQA PAST PAPERS WITH ANSWERS

ADVANCED HIGHER

CHEMISTRY
2006-2009

SQA

BrightRED
PUBLISHING

First exam published in 2006.

Published by Bright Red Publishing Ltd, 6 Stafford Street, Edinburgh EH3 7AU

tel: 0131 220 5804 fax: 0131 220 6710 info@brightredpublishing.co.uk www.brightredpublishing.co.uk

ISBN 978-1-84948-073-4

A CIP Catalogue record for this book is available from the British Library.

Bright Red Publishing is grateful to the copyright holders, as credited on the final page of the book, for permission to use their material.

Every effort has been made to trace the copyright holders and to obtain their permission for the use of copyright material.

Bright Red Publishing will be happy to receive information allowing us to rectify any error or omission in future editions.

ADVANCED HIGHER

2006

[BLANK PAGE]

X012/701

NATIONAL
QUALIFICATIONS
2006

TUESDAY, 30 MAY
9.00 AM – 11.30 AM

CHEMISTRY
ADVANCED HIGHER

Reference may be made to the Chemistry Higher and Advanced Higher Data Booklet (1999 edition).

SECTION A – 40 marks

Instructions for completion of **SECTION A** are given on page two.

For this section of the examination you must use an **HB pencil**.

SECTION B – 60 marks

All questions should be attempted.

Answers must be written clearly and legibly in ink.

SCOTTISH
QUALIFICATIONS
AUTHORITY

SECTION A

Read carefully

1 Check that the answer sheet provided is for **Chemistry Advanced Higher (Section A)**.

2 For this section of the examination you must use an **HB pencil** and, where necessary, an eraser.

3 Check that the answer sheet you have been given has **your name**, **date of birth**, **SCN** (Scottish Candidate Number) and **Centre Name** printed on it.

   Do not change any of these details.

4 If any of this information is wrong, tell the Invigilator immediately.

5 If this information is correct, **print** your name and seat number in the boxes provided.

6 The answer to each question is **either** A, B, C or D. Decide what your answer is, then, using your pencil, put a horizontal line in the space provided (see sample question below).

7 There is **only one correct** answer to each question.

8 Any rough working should be done on the question paper or the rough working sheet, **not** on your answer sheet.

9 At the end of the exam, put the **answer sheet for Section A inside the front cover of your answer book**.

Sample Question

To show that the ink in a ball-pen consists of a mixture of dyes, the method of separation would be

        A chromatography

        B fractional distillation

        C fractional crystallisation

        D filtration.

The correct answer is **A**—chromatography. The answer **A** has been clearly marked in **pencil** with a horizontal line (see below).

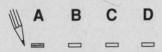

Changing an answer

If you decide to change your answer, carefully erase your first answer and using your pencil, fill in the answer you want. The answer below has been changed to **D**.

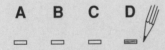

1. Sodium vapour street lamps emit yellow light because

   A sodium vapour is burning and giving out a yellow glow

   B sodium vapour filters out all the light from the filament except yellow

   C energy corresponding to yellow light is given out as electrons in sodium move to higher energies

   D energy corresponding to yellow light is given out as electrons in sodium move to lower energies.

2. What is the change in the three-dimensional arrangement of the bonds round the B atom in the following reaction?

   $$BF_4^- \rightarrow BF_3 + F^-$$

   A Square planar to trigonal planar

   B Tetrahedral to trigonal planar

   C Tetrahedral to pyramidal

   D Square planar to pyramidal

3. Which of the following oxides is amphoteric?

   A Aluminium oxide

   B Calcium oxide

   C Copper(II) oxide

   D Sulphur dioxide

4. The electrical conductivity of semiconductors

   A decreases with increasing temperature

   B increases with increasing temperature

   C decreases on exposure to light

   D increases with the removal of dopant atoms.

5. Which of the following molecules has the greatest number of non-bonding electron pairs (lone pairs)?

   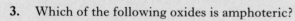

   A   H—C—Cl   (with H above and H below C)

   B   H—C—O—H   (with H above and H below C)

   C   H—C—N   (with H above and H below C, and two H on N)

   D   H—C=O   (with H above C)

6. A white solid gives a lilac flame colour. It reacts with water releasing hydrogen gas and forming a strongly alkaline solution.

   The solid could be

   A calcium oxide

   B potassium oxide

   C calcium hydride

   D potassium hydride.

7. Hund's rule states that

   A it is impossible to define both the position and momentum of an electron simultaneously

   B electrons occupy orbitals in order of increasing energy

   C electrons occupy degenerate orbitals singly with parallel spins before spin pairing occurs

   D the energy of an electron in an atom is quantised.

   [Turn over

Questions 8 and 9 refer to the analysis of a salt whose formula is $Pt(NH_3)_xCl_y$.

8. 0.02 moles of this salt required $40.0\,cm^3$ of $2.0\,mol\,l^{-1}$ nitric acid for exact neutralisation.

The number of moles of NH_3 per mole of salt is

A 2

B 4

C 6

D 8.

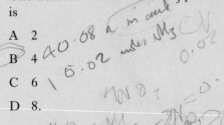

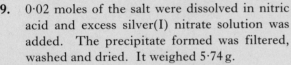

9. 0.02 moles of the salt were dissolved in nitric acid and excess silver(I) nitrate solution was added. The precipitate formed was filtered, washed and dried. It weighed $5.74\,g$.

The number of moles of chloride ions per mole of the salt is

A 1

B 2

C 3

D 4.

10. One mole of barium chloride ($BaCl_2$) contains

A 1 mole of positive ions

B 1 mole of molecules

C 2 moles of atoms

D 2 moles of ions.

11. Which change in reaction conditions will shift the position of equilibrium to the right in this reaction?

$$2SO_2(g) + O_2(g) \rightleftharpoons 2SO_3(g) \quad \Delta H^\circ = -197\,kJ\,mol^{-1}$$

A Increasing the temperature

B Removal of some oxygen gas

C Increasing the pressure

D Adding a catalyst

12. An organic acid, **X**, was dissolved in water and then shaken with ethoxyethane until equilibrium was established.

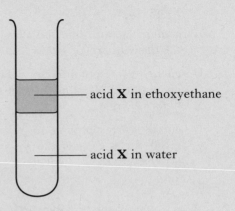

— acid **X** in ethoxyethane

— acid **X** in water

The value of the partition coefficient for this system will be altered by changing the

A temperature

B volume of water

C original concentration of acid **X**

D original mass of acid **X**.

13. The pH of a buffer prepared by mixing equal volumes of $0.1\,mol\,l^{-1}$ ethanoic acid and $0.2\,mol\,l^{-1}$ sodium ethanoate is

A 2.1

B 2.7

C 4.5

D 5.1.

14. Ethanoic acid is a weak acid and hydrochloric acid is a strong acid.

Which of the following is **not** correct?

A The pH of $0.1\,mol\,l^{-1}$ hydrochloric acid is 1.

B $20.0\,cm^3$ of $0.1\,mol\,l^{-1}$ sodium hydroxide is exactly neutralised by $20.0\,cm^3$ of $0.1\,mol\,l^{-1}$ ethanoic acid.

C The pH of $0.1\,mol\,l^{-1}$ hydrochloric acid is lower than that of $0.1\,mol\,l^{-1}$ ethanoic acid.

D The K_a value of ethanoic acid is greater than that of hydrochloric acid.

15. The use of an indicator is **not** appropriate in titrations involving

   A hydrochloric acid solution and methylamine solution

   B nitric acid solution and potassium hydroxide solution

   C methanoic acid solution and ammonia solution

   D propanoic acid solution and sodium hydroxide solution.

16. In the presence of bright light, hydrogen and chlorine react explosively. One step in the reaction is shown below.

$$H_2(g) + Cl(g) \rightarrow HCl(g) + H(g)$$

   Using information from the Data Booklet, the enthalpy change, in $kJ\ mol^{-1}$, for this step is calculated as

   A -189

   B -4

   C $+4$

   D $+189$.

17. For which of the following reactions does the $\Delta H°$ value correspond to **both** the enthalpy of combustion of an element and the enthalpy of formation of a compound?

   A $C(s) + \frac{1}{2}O_2(g) \rightarrow CO(g)$

   B $H_2(g) + O_2(g) \rightarrow H_2O_2(\ell)$

   C $2Na(s) + \frac{1}{2}O_2(g) \rightarrow Na_2O(s)$

   D $Mg(s) + \frac{1}{2}O_2(g) \rightarrow MgO(s)$

18. In which of the following reactions would the energy change represent the lattice enthalpy of sodium chloride?

   A $Na^+(g) + Cl^-(g) \rightarrow NaCl(s)$

   B $Na(g) + Cl(g) \rightarrow NaCl(s)$

   C $Na(s) + \frac{1}{2}Cl_2(g) \rightarrow NaCl(s)$

   D $Na(s) + Cl(g) \rightarrow NaCl(s)$

19. A Born-Haber cycle can be used to calculate the lattice enthalpy of sodium chloride. Which of the following is **not** required in the calculation?

   A The bond enthalpy of chlorine

   B The first ionisation energy of chlorine

   C The first ionisation energy of sodium

   D The enthalpy of formation of sodium chloride

20. Which of the following reactions results in a **decrease** in entropy?

   A $N_2(g) + 3H_2(g) \rightarrow 2NH_3(g)$

   B $N_2O_4(g) \rightarrow 2NO_2(g)$

   C $CaCO_3(s) \rightarrow CaO(s) + CO_2(g)$

   D $C(s) + H_2O(g) \rightarrow CO(g) + H_2(g)$

21. The equilibrium constant for a particular hydrolysis reaction has the value 3×10^4 at $25\,°C$. From this we can conclude that, at $25\,°C$, this hydrolysis reaction is

   A fast

   B feasible

   C exothermic

   D endothermic.

[Turn over

22.

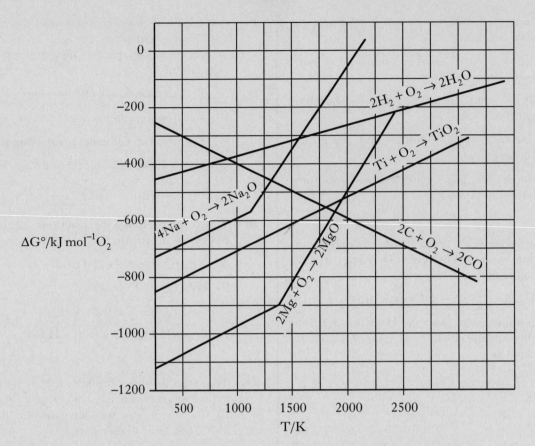

$\Delta G^\circ / \text{kJ mol}^{-1} O_2$

The reduction of TiO_2 to Ti is thermodynamically feasible at 1500 K using

A   hydrogen

B   magnesium

C   carbon

D   sodium.

23.

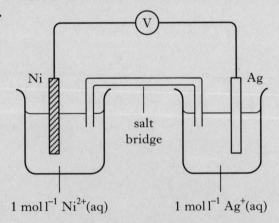

1 mol l^{-1} Ni^{2+}(aq)   1 mol l^{-1} Ag^{+}(aq)

In the cell, which of the following is reduced?

(Assume standard conditions.)

A   Ni(s)

B   Ni^{2+}(aq)

C   Ag(s)

D   Ag^{+}(aq)

24.   For the reaction

$$2NO(g) + Cl_2(g) \rightarrow 2NOCl(g)$$

the suggested mechanism is

$$NO(g) + Cl_2(g) \xrightarrow{\text{slow}} NOCl_2(g)$$

$$NOCl_2(g) + NO(g) \xrightarrow{\text{fast}} 2NOCl(g)$$

The rate equation is

A   rate = $k[NO][Cl_2]$

B   rate = $k[NO]^2[Cl_2]$

C   rate = $k[NOCl_2][NO]$

D   rate = $k[NO_2]^2[NOCl_2][Cl_2]$.

25. In a series of experiments, P and Q reacted to form R. The times taken to produce a fixed concentration of R were recorded.

Experiment	Initial [P] /mol l^{-1}	Initial [Q] /mol l^{-1}	Time/s
1	0·05	0·05	46
2	0·05	0·10	23
3	0·10	0·05	46

The rate equation for this reaction is

A   rate = k[P]

B   rate = k[Q]

C   rate = k[Q]^2

D   rate = k[P][Q].

26. Which line in the table correctly describes the types of reaction in the following sequence?

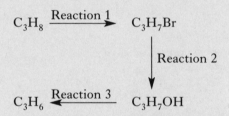

	Reaction 1	Reaction 2	Reaction 3
A	addition	substitution	dehydration
B	addition	addition	condensation
C	substitution	substitution	dehydration
D	substitution	addition	condensation

27. When ethene reacts with bromine in the presence of potassium chloride, CH_2BrCH_2Cl and CH_2BrCH_2Br are both formed. The first of these two compounds is produced because chloride ions

A   compete with ethene to form an ionic intermediate

B   compete with bromide ions in attacking a cyclic ion intermediate

C   attack the CH_2BrCH_2Br which had originally formed, displacing the less reactive bromide ions

D   react with bromine to give chlorine which then attacks the ethene.

28. Which line in the table has the correct number and type of bonds in

$$H - C \equiv C - C = C \overset{H}{\underset{H}{<}}$$

	Number of σ-bonds	Number of π-bonds
A	7	3
B	7	2
C	5	2
D	5	5

29. Which of the following best describes the bonding in alkanes?

A   sp^2 hybridisation of the carbon atoms giving sigma bonds only

B   sp^2 hybridisation of the carbon atoms giving sigma and pi bonds

C   sp^3 hybridisation of the carbon atoms giving sigma bonds only

D   sp^3 hybridisation of the carbon atoms giving sigma and pi bonds

30. Which of the following represents a termination step in a chain reaction?

A $Cl\bullet + Cl\bullet \rightarrow Cl_2$

B $Cl\bullet + CH_4 \rightarrow CH_3\bullet + HCl$

C $CH_3\bullet + Cl_2 \rightarrow CH_3Cl + Cl\bullet$

D $Cl_2 \rightarrow Cl\bullet + Cl\bullet$

31. Which one of the following statements about ethoxyethane is **not** correct?

A   It burns readily in air.

B   It is isomeric with butan-1-ol.

C   It has a higher boiling point than butan-1-ol.

D   It may be prepared from sodium ethoxide and bromoethane.

[Turn over

32. Which of the following, when reacted with ethane-1,2-diol, would form a polyester?

A   HO — C(=O) — ⟨benzene ring⟩

B   HO — C(=O) — ⟨benzene ring⟩ — C(=O) — OH

C   HO — C(=O) — ⟨benzene ring⟩ — C(=O) — H

D   H — C(=O) — ⟨benzene ring⟩ — C(=O) — H

33. Which of the following can be oxidised by Tollens' reagent?

A CH_3COCH_3

B CH_3CHO

C CH_3OH

D $(CH_3)_2CHOH$

34. Which of the following is the strongest base?

A CH_3CH_2OH

B   ⟨benzene ring⟩— OH

C $CH_3CH_2NH_2$

D   ⟨benzene ring⟩— NH_2

35. Naphthalene, $C_{10}H_8$, has the following structure.

The number of moles of hydrogen gas required for the complete hydrogenation of 12·8 g naphthalene will be

A   0·1

B   0·4

C   0·5

D   0·8.

36. Which of the following compounds has a geometric isomer?

A
$$\begin{array}{ccc} & H & Cl \\ & | & | \\ H - & C - & C - H \\ & | & | \\ & Cl & H \end{array}$$

B
$$\begin{array}{cc} H & Cl \\ \diagdown & \diagup \\ C & = C \\ \diagup & \diagdown \\ H & H \end{array}$$

C
$$\begin{array}{cc} H & Cl \\ \diagdown & \diagup \\ C & = C \\ \diagup & \diagdown \\ H & Cl \end{array}$$

D
$$\begin{array}{cc} H & Cl \\ \diagdown & \diagup \\ C & = C \\ \diagup & \diagdown \\ Cl & H \end{array}$$

37. An analysis of an organic compound found in meteorite rocks shows the following percentage composition by mass.

C = 37·5%   H = 12·5%   O = 50%

The empirical (simplest) formula for the compound is

A CH_4O

B C_3HO_4

C $C_3H_{12}O_3$

D CH_2O_2.

38. A simplified mass spectrum of an organic compound is shown.

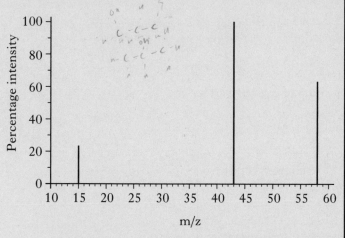

Which of the following compounds gave this spectrum?

A   Propane

B   Propan-1-ol

C   Propan-2-ol

D   Propanone

39. Which of the following analytical techniques depends on the vibrations within molecules?

A   Colorimetry

B   Atomic emission spectroscopy

C   Infra-red absorption spectroscopy

D   Mass spectroscopy

40. Salbutamol is used to treat asthma. It behaves like the body's natural active compound by binding to receptors on the muscles of the air passages. This relaxes the muscles and gives relief from breathing difficulties. Salbutamol is

A   an agonist

B   an antagonist

C   a pharmacaphore

D   a receptor.

[END OF SECTION A]

Candidates are reminded that the answer sheet for Section A MUST be placed INSIDE the front cover of your answer book.

[Turn over

SECTION B

All answers must be written clearly and legibly in ink.

Marks

1. Molten iron, made in a blast furnace, often contains sulphur and phosphorus impurities which must be removed.

   Bubbling carbon dioxide gas through molten iron removes the sulphur.

   The carbon dioxide gas is produced by the decomposition of calcium carbonate.

   $$CaCO_3(s) \longrightarrow CaO(s) + CO_2(g)$$

   (a)

Substance	Standard enthalpy of formation, $\Delta H°/kJ\,mol^{-1}$	Standard entropy, $S°/J\,K^{-1}\,mol^{-1}$
CO_2	−393·5	213·8
$CaCO_3$	−1206·9	92·9
CaO	−635·1	38·1

   For the decomposition of calcium carbonate, use the data in the table to calculate:

   (i)   the standard enthalpy change, $\Delta H°$, in $kJ\,mol^{-1}$;   1

   (ii) the standard entropy change, $\Delta S°$, in $J\,K^{-1}\,mol^{-1}$;   1

   (iii) the theoretical temperature at which the reaction just becomes feasible.   2

   (b) The phosphorus impurity is removed by first converting it into the oxide, P_4O_{10}. This oxide then reacts with calcium oxide to produce a useful fertiliser.

   (i)   Calculate the oxidation number of phosphorus in P_4O_{10}.   1

   (ii) Name the fertiliser formed.   1

   (6)

2. $25·0\,cm^3$ of an acidified solution of potassium oxalate, $K_2C_2O_4$, was heated to $80\,°C$ and titrated with a standard solution of $0·020\,mol\,l^{-1}$ potassium permanganate, $KMnO_4$.

   The end-point was reached when $22·5\,cm^3$ of $KMnO_4$ solution had been added.

   The ion-electron equations for the reactions involved are:

   $$C_2O_4{}^{2-}(aq) \longrightarrow 2CO_2(g) + 2e^-$$
   $$MnO_4{}^-(aq) + 8H^+(aq) + 5e^- \longrightarrow Mn^{2+}(aq) + 4H_2O(\ell)$$

   (a) How would the end-point of the titration be determined?   1

   (b) Write the redox equation for the reaction.   1

   (c) Calculate the concentration of the potassium oxalate solution used in this titration.   3

   (5)

Marks

3. The table shows two quantum numbers for the 10 electrons in a neon atom.

Electron	First Quantum Number (n)	Second Quantum Number (ℓ)
1	1	0
2	1	0
3	2	0
4	2	0
5	2	1
6	2	1
7	2	1
8	2	1
9	2	1
10	2	1

(a) Write the electronic configuration for a neon atom in terms of s and p orbitals. $1s^2 2s^2 2p^6$

1

(b) Electrons 5 to 10 can be described as degenerate.

What is meant by the term "degenerate"? *equal energy*

1

(c) The second quantum number, ℓ, is related to the shape of the orbitals.

Draw the shape of an orbital when $\ell = 1$.

1

(d) What are the first and second quantum numbers for the outer electron in a sodium atom?

1

(4)

[Turn over

Marks

4. In the stratosphere, oxygen molecules absorb ultraviolet radiation and break up to form oxygen atoms.

$$O=O \longrightarrow O + O$$

(a) The bond enthalpy of $O=O$ is $497 \, kJ \, mol^{-1}$. Calculate the wavelength, in nm, of the ultraviolet radiation required to break up 1 mole of oxygen molecules into oxygen atoms.   **3**

(b) Some of the oxygen atoms react with oxygen molecules to produce ozone, O_3.

Reaction 1 $O + O_2 \longrightarrow O_3$ $\Delta H^{\circ} = -106 \, kJ \, mol^{-1}$

Other oxygen atoms react to produce oxygen molecules.

Reaction 2 $O + O \longrightarrow O_2$ $\Delta H^{\circ} = -497 \, kJ \, mol^{-1}$

Oxygen atoms can also react with ozone to produce oxygen molecules.

Reaction 3 $O + O_3 \longrightarrow O_2 + O_2$

(i) Use Hess's law to calculate the standard enthalpy change, in $kJ \, mol^{-1}$, for **Reaction 3**.   **1**

(ii) A Lewis dot diagram for an oxygen molecule is

Draw a similar diagram for an ozone molecule.   **1**

    (5)

5. Solutions of NaH_2PO_4 are acidic because the $H_2PO_4^-$ ion partially dissociates.

$$H_2PO_4^-(aq) \rightleftharpoons H^+(aq) + HPO_4^{2-}(aq) \quad pK_a = 7{\cdot}2$$

(a) Write the expression for the acid dissociation constant, K_a.   **1**

(b) Calculate the pH of $0{\cdot}1 \, mol \, l^{-1} \, NaH_2PO_4$ solution.   **2**

(c) NaH_2PO_4 is used with $NaHCO_3$ in baking powders, to produce carbon dioxide.

$$H_2PO_4^-(aq) + HCO_3^-(aq) \rightleftharpoons HPO_4^{2-}(aq) + H_2O(\ell) + CO_2(g)$$

Explain how HCO_3^- acts as a base in this reaction.   **1**

    (4)

Marks

6. Most cars are powered by the combustion of petrol which consists mainly of octane isomers. Currently research is being carried out on replacing petrol with hydrogen as a fuel.

(*a*)

Fuel	Combustion Reaction	$\Delta H^\circ / kJ\,mol^{-1}$
Hydrogen	$H_2 + \tfrac{1}{2}O_2 \longrightarrow H_2O$	-286
Petrol	$C_8H_{18} + 12\tfrac{1}{2}O_2 \longrightarrow 8CO_2 + 9H_2O$	-5100

Calculate the energy produced per gram of each fuel.   **1**

(*b*) Research is also being carried out on replacing engines with fuel cells such as the direct methanol fuel cell shown.

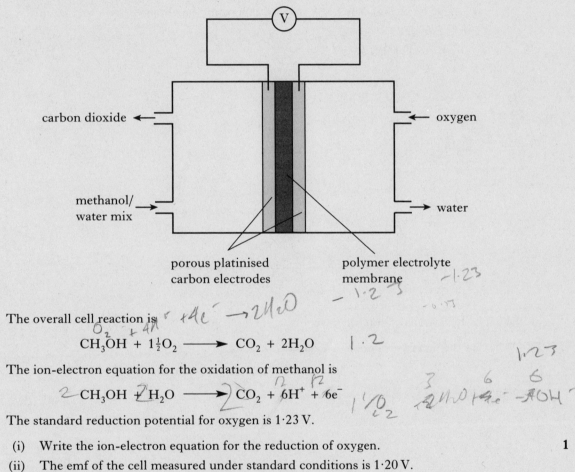

The overall cell reaction is

$$CH_3OH + 1\tfrac{1}{2}O_2 \longrightarrow CO_2 + 2H_2O$$

The ion-electron equation for the oxidation of methanol is

$$CH_3OH + H_2O \longrightarrow CO_2 + 6H^+ + 6e^-$$

The standard reduction potential for oxygen is 1·23 V.

  (i) Write the ion-electron equation for the reduction of oxygen.   **1**

  (ii) The emf of the cell measured under standard conditions is 1·20 V.

    Calculate the standard **reduction** potential for the ion-electron reaction involving the methanol.   **1**

  (iii) Calculate the standard free energy change, ΔG°, in kJ per mole of methanol, for the cell reaction.   **3**

(6)

[Turn over

Marks

7. In a **PPA**, the kinetics of the acid-catalysed propanone/iodine reaction were studied.

$$CH_3COCH_3(aq) + I_2(aq) \xrightarrow{H^+(aq)} CH_3COCH_2I(aq) + HI(aq)$$

The reaction is first order with respect to propanone and first order with respect to the hydrogen ions which catalyse the reaction. The order with respect to iodine is unknown. The rate equation is

$$Rate = k[I_2]^x[CH_3COCH_3][H^+]$$

The aim of the experiment was to determine **x**.

(a) How did the initial concentrations of the propanone and acid compare with that of the iodine to allow the value of **x** to be determined?   1

(b) The experiment proved that the order of the reaction with respect to iodine was zero. Copy the axes shown and sketch the graph which would be obtained.

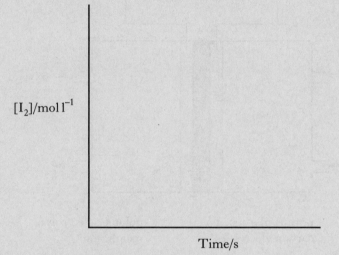

1

(c) What is the overall order of the reaction?   1

(d) What are the units for the rate constant, k?   1

(4)

Marks

8. In a **PPA**, cyclohexene was prepared from cyclohexanol by dehydration.

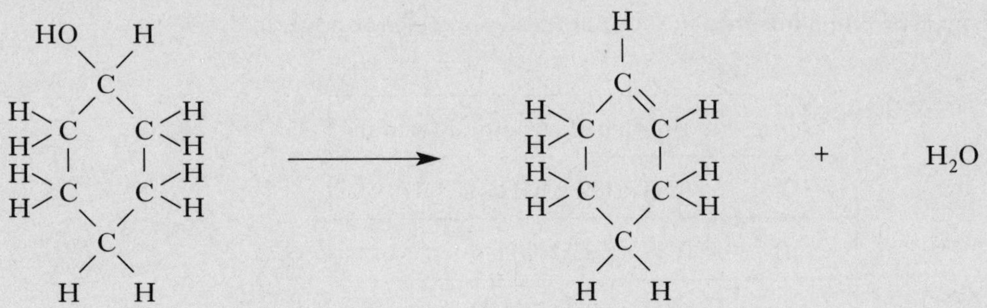

(a) Which reagent was used to convert cyclohexanol to cyclohexene?   **1**

(b) Distillation was used to separate the cyclohexene product from the reaction mixture because cyclohexanol has a higher boiling point than cyclohexene.

Explain why cyclohexanol has the much higher boiling point.   **1**

(c) To purify the cyclohexene distillate, a saturated sodium chloride solution was added to it in a separating funnel.

Why was sodium chloride solution used instead of water?   **1**

(d) Cyclohexanol can also be oxidised to the cycloketone, cyclohexanone.

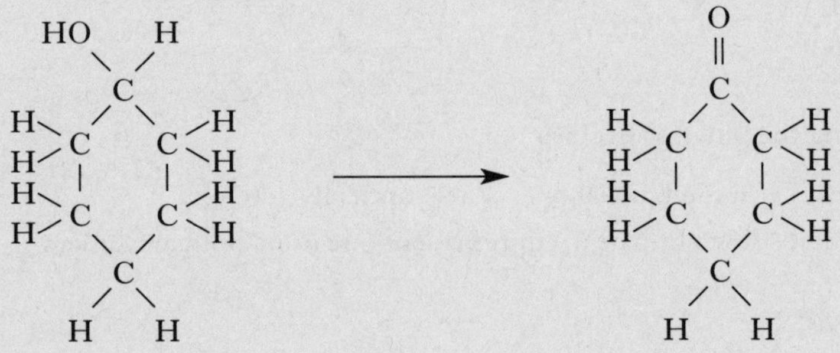

Ketones can be identified using 2,4-dinitrophenylhydrazine solution (Brady's reagent).

How would this be used to identify the ketone as cyclohexanone?   **2**

(5)

[Turn over

Marks

9. Mixtures of the isomers of the alcohol, $C_5H_{11}OH$, are used as solvents for resins and oily materials.

   The shortened structural formulae for four of these isomers are shown in the table.

Isomer	Shortened structural formula
A	$(CH_3)(C_2H_5)CHCH_2OH$
B	$(CH_3)_3CCH_2OH$
C	$(CH_3)_2(C_2H_5)COH$
D	$(C_2H_5)_2CHOH$

   (a) Which isomer is the tertiary alcohol?   1

   (b) Another isomer of $C_5H_{11}OH$ displays optical isomerism.

   One of its optical isomers is shown below.

   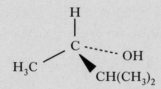

   Draw a diagram representing the other optical isomer.   1

   (c) One of the four isomers, A to D, in the table above, is also optically active.

   Draw a similar diagram to that shown in part (b) to represent one of its optical isomers.   1

      (3)

10. Fusel alcohols are components of fusel oil which is obtained during the process of brewing beer. They contribute to the flavour of the beer and are also the major cause of hangovers. The most important fusel alcohols are

    3-methylbutan-1-ol
    2-methylbutan-1-ol
    propan-1-ol.

    (a) Give a reason why propan-1-ol is the most soluble of these alcohols in water.   1

    (b) Other flavours found in beer are caused by esters.

    Esters can be formed by reacting alcohols with carboxylic acids.

       (i) What can be used in place of carboxylic acids to form esters?   1

       (ii) What advantage is there in using this type of reagent?   1

       (3)

Marks

11. Benzene is one of the most important aromatic feedstocks in the chemical industry.

    Four electrophilic substitution reactions which benzene undergoes are shown.

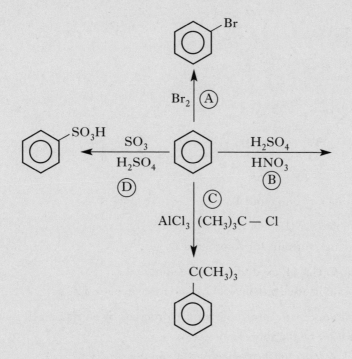

(a) (i) Which catalyst is required to carry out reaction (A)?   1

    (ii) What is the organic product in reaction (B)?   1

    (iii) The specific name for reaction (C) is alkylation.

        What is the specific name for reaction (D)?   1

(b) Both benzene and graphite have delocalised electrons.

    Suggest why benzene does not conduct electricity.   1

    (4)

[Turn over

Marks

12. The diagram illustrates two methods of preparing compound **Z**.

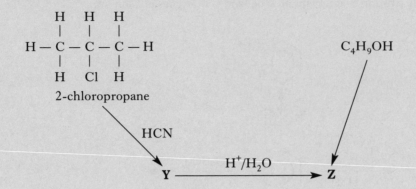

(a) Draw a structural formula for compound **Y**.   1

(b) Compound **Z** has the molecular formula $C_4H_8O_2$.
Name **and** draw a structural formula for compound **Z**.   2

(c) (i) Name the alcohol, C_4H_9OH, used to prepare compound **Z**.   1

(ii) Which reagent could be used to convert C_4H_9OH to compound **Z**?   1

(d) The conversion of 2-chloropropane into compound **Y** proceeds by an S_N2 mechanism.

(i) Explain what is meant by the abbreviation S_N2.   2

(ii) Draw a structural formula for the transition state in this reaction.   1

(8)

Marks

13. A proton nmr spectrum was produced for ethanal.

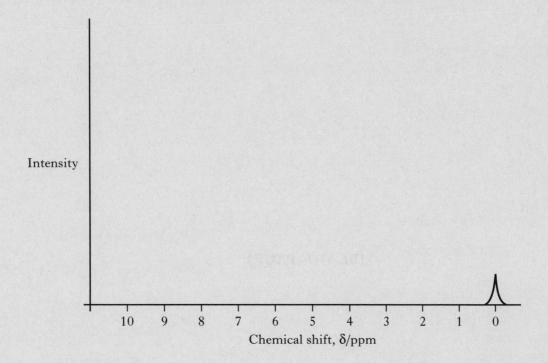

ethanal

(a) Copy and complete the following diagram to show the approximate positions and relative heights of the two peaks in the proton nmr spectrum of ethanal.

2

(b) Which reference substance is used in proton nmr spectroscopy and causes the peak at $\delta = 0$ ppm?

1

(3)

[END OF QUESTION PAPER]

[BLANK PAGE]

2007

[BLANK PAGE]

X012/701

| NATIONAL QUALIFICATIONS 2007 | TUESDAY, 29 MAY 9.00 AM – 11.30 AM | CHEMISTRY ADVANCED HIGHER |

Reference may be made to the Chemistry Higher and Advanced Higher Data Booklet (2007 edition).

SECTION A – 40 marks

Instructions for completion of **SECTION A** are given on page two.

For this section of the examination you must use an **HB pencil**.

SECTION B – 60 marks

All questions should be attempted.

Answers must be written clearly and legibly in ink.

SCOTTISH QUALIFICATIONS AUTHORITY

SECTION A

Read carefully

1 Check that the answer sheet provided is for **Chemistry Advanced Higher (Section A)**.

2 For this section of the examination you must use an **HB pencil** and, where necessary, an eraser.

3 Check that the answer sheet you have been given has **your name, date of birth, SCN** (Scottish Candidate Number) and **Centre Name** printed on it.

   Do not change any of these details.

4 If any of this information is wrong, tell the Invigilator immediately.

5 If this information is correct, **print** your name and seat number in the boxes provided.

6 The answer to each question is **either** A, B, C or D. Decide what your answer is, then, using your pencil, put a horizontal line in the space provided (see sample question below).

7 There is **only one correct** answer to each question.

8 Any rough working should be done on the question paper or the rough working sheet, **not** on your answer sheet.

9 At the end of the exam, put the **answer sheet for Section A inside the front cover of your answer book**.

Sample Question

To show that the ink in a ball-pen consists of a mixture of dyes, the method of separation would be

    A chromatography

    B fractional distillation

    C fractional crystallisation

    D filtration.

The correct answer is **A**—chromatography. The answer **A** has been clearly marked in **pencil** with a horizontal line (see below).

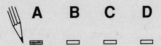

Changing an answer

If you decide to change your answer, carefully erase your first answer and using your pencil, fill in the answer you want. The answer below has been changed to **D**.

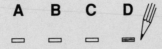

1. What type of bonding exists in an element which forms a gaseous oxide?

   A Ionic

   B Covalent (polar)

   C Covalent (non-polar)

   D Metallic

2. Which of the following compounds is likely to have the **most** covalent character?

   A Tin(IV) iodide

   B Iron(II) chloride

   C Lithium fluoride

   D Potassium bromide

3. Which of the following is **not** a form of electromagnetic radiation?

   A Beta radiation

   B Gamma radiation

   C Ultra-violet radiation

   D Infra-red radiation

4. **All** noble gases are characterised in terms of electrons by the completion of the outermost orbital. This orbital is

   A an s-orbital

   B a p-orbital

   C a d-orbital

   D an s or p-orbital.

5. Infra-red radiation can be used in the analysis and identification of organic compounds. Compared to visible radiation, infra-red radiation has a

   A shorter wavelength and higher frequency

   B longer wavelength and lower velocity

   C longer wavelength and lower frequency

   D shorter wavelength and higher velocity.

6. When an ammonia molecule accepts a proton to form an ammonium ion, there is a change of shape from

   A pyramidal to tetrahedral

   B pyramidal to trigonal planar

   C trigonal planar to pyramidal

   D trigonal planar to tetrahedral.

7. Isoelectronic species are molecules or polyatomic ions with the same total number of electrons.

   Which of the following contains two isoelectronic species?

   A NH_4^+ and CH_4

   B NH_4^+ and BF_3

   C BF_3 and NH_3

   D PF_5 and BF_4^-

8. A superconductor is a material

   A whose electrical conductivity decreases with decreasing temperature

   B that does not conduct electricity unless doped with another material

   C that can conduct electricity with zero resistance

   D whose electrical conductivity increases with increasing temperature.

9. The statement that "an orbital can accommodate, at most, two electrons, and if so, they must be of opposite spin" is based on

   A Hund's rule

   B the aufbau principle

   C the Pauli exclusion principle

   D Heisenberg's uncertainty principle.

[Turn over

10.

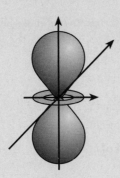

The above is a diagrammatic representation of the shape of

A   any p-orbital

B   a specific p-orbital

C   any d-orbital

D   a specific d-orbital.

11. An ionic compound is thought to contain hydride ions. One way to show this could be to

A   electrolyse the molten compound and test for hydrogen gas at the negative electrode

B   electrolyse the molten compound and test for hydrogen gas at the positive electrode

C   dissolve the compound in water and test for an acidic pH

D   dissolve the compound in water and test for an alkaline pH.

12.

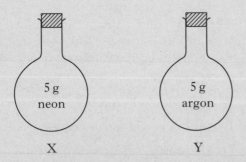

The number of atoms of gas in flask Y is approximately

A   equal to the number of atoms of gas in flask X

B   twice the number of atoms of gas in flask X

C   one half the number of atoms of gas in flask X

D   one quarter the number of atoms of gas in flask X.

13. In a gravimetric analysis of silver, a precipitate of silver(I) chromate was produced by adding excess potassium chromate to a solution containing silver(I) ions.

If $5 \cdot 795$ g of Ag_2CrO_4 was produced, the mass of silver in the solution was

A $1 \cdot 884$ g

B $3 \cdot 318$ g

C $3 \cdot 769$ g

D $8 \cdot 910$ g.

14. $N_2O_4(g) \rightleftharpoons 2NO_2(g)$ $\Delta H^\circ = +57\,kJ\,mol^{-1}$

Which of the following will increase the equilibrium constant for the reaction?

A   Use of a catalyst

B   Increase of pressure

C   Increase of temperature

D   Decrease of temperature

15. An aqueous solution of an organic acid, **X**, was shaken with ethoxyethane until the following equilibrium was established.

$$\mathbf{X}(\text{water}) \rightleftharpoons \mathbf{X}(\text{ethoxyethane})$$

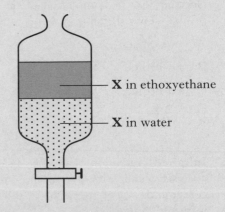

$20 \cdot 0\,cm^3$ of the upper layer needed $15 \cdot 0\,cm^3$ of $0 \cdot 010\,mol\,l^{-1}$ NaOH(aq) for neutralisation.

$20 \cdot 0\,cm^3$ of the lower layer needed $12 \cdot 0\,cm^3$ of $0 \cdot 010\,mol\,l^{-1}$ NaOH(aq) for neutralisation.

The value of the partition coefficient is

A $0 \cdot 80$

B $1 \cdot 25$

C $1 \cdot 33$

D $1 \cdot 67$.

16.

$$2Fe(s) + \tfrac{3}{2}O_2(g) \rightarrow Fe_2O_3(s) \quad \Delta H^\circ = -822\,kJ\,mol^{-1}$$
$$C(s) + O_2(g) \rightarrow CO_2(g) \quad \Delta H^\circ = -394\,kJ\,mol^{-1}$$

Which of the following is the standard enthalpy change, in $kJ\,mol^{-1}$, for the reaction shown below?

$$Fe_2O_3(s) + \tfrac{3}{2}C(s) \rightarrow 2Fe(s) + \tfrac{3}{2}CO_2(g)$$

A +231

B +428

C +1216

D +1413

17. Which of the following can **not** be determined by a single experiment? The enthalpy of

A formation of ethyne

B combustion of methanol

C solution of sodium hydroxide

D neutralisation of hydrochloric acid by sodium hydroxide.

18. Consider the following bond enthalpies.

Bond	Enthalpy/$kJ\,mol^{-1}$
Br — Br	194
H — Br	362
C — H	414
C — Br	285

What is the enthalpy change, in $kJ\,mol^{-1}$, for the following reaction?

```
    H   H
    |   |
H — C — C — H   +   Br — Br
    |   |
    H   H

            ↓

    H   H
    |   |
H — C — C — Br   +   H — Br
    |   |
    H   H
```

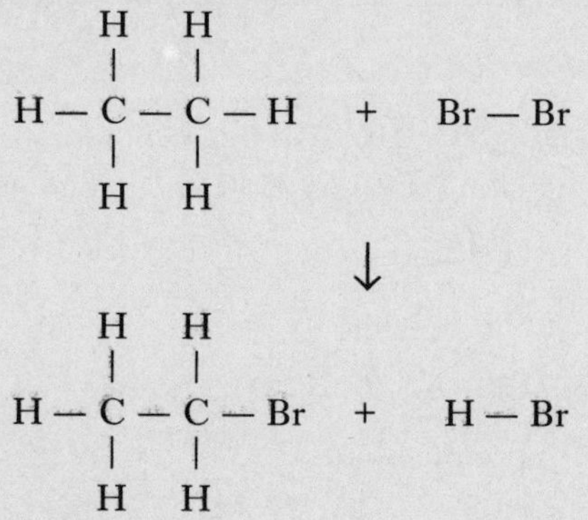

A −1255

B −39

C +39

D +1255

19. Which of the following is exothermic?

A $K(s) \rightarrow K(g)$

B $K(g) \rightarrow K^+(g) + e^-$

C $\tfrac{1}{2}Br_2(\ell) \rightarrow Br(g)$

D $Br(g) + e^- \rightarrow Br^-(g)$

20. Which of the following is likely to have the lowest standard entropy value at $100\,^\circ C$?

A Neon

B Mercury

C Sulphur

D Phosphorus

21. An Ellingham diagram shows how ΔG° for a chemical reaction varies with

A ΔH°

B ΔS°

C time

D temperature.

22. Which of the following reactions must be exothermic? One in which

A ΔG° is negative

B ΔS° is positive

C both ΔG° and ΔS° are negative

D both ΔG° and ΔS° are positive.

23. In the electrochemical cell

$$Cr(s)\,Cr^{3+}(aq)\,Ag^+(aq)\,Ag(s)$$

operating under standard conditions, which of the following is **not** true?

A The emf is $1 \cdot 54\,V$.

B The silver electrode decreases in mass.

C Oxidation takes place at the chromium electrode.

D Electrons flow from chromium to silver in the external circuit.

[Turn over

24. The following reaction is first order with respect to each of the reactants.

$$A + B \rightarrow C + D$$

Which of the following is correct?

A The rate of the reaction is independent of the concentration of either A or B.

B The overall reaction is first order.

C If the initial concentrations of A and B are both doubled, the rate of the reaction will be doubled.

D As the reaction proceeds, its rate will decrease.

25. Which of the following does **not** apply to the reaction between methane and chlorine?

It involves

A homolytic fission

B a chain reaction

C a carbocation intermediate

D free radicals.

26. Which of the following amines has the lowest boiling point?

A $C_4H_9NH_2$

B $C_3H_7NHCH_3$

C $C_2H_5NHC_2H_5$

D $C_2H_5N(CH_3)_2$

27. Chlorobenzene, nitrobenzene and ethylbenzene can all be formed from benzene by

A electrophilic substitution

B electrophilic addition

C nucleophilic substitution

D nucleophilic addition.

28.

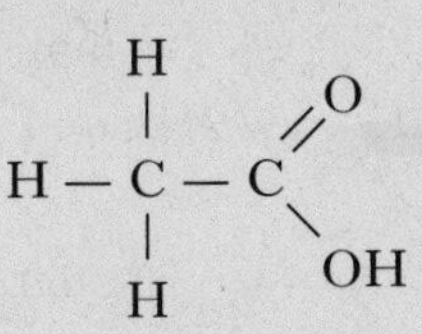

X 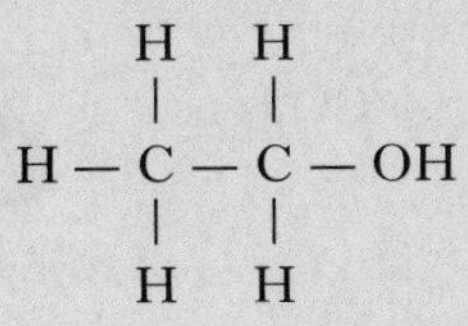   Y 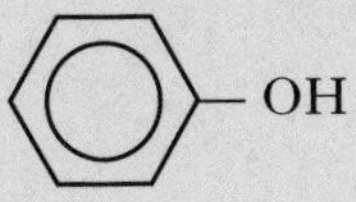   Z

Which of the following shows the above compounds in order of **increasing** acid strength?

A X Y Z

B Y Z X

C Z X Y

D X Z Y

29. Which of the following has nucleophilic properties?

A Na

B Br^+

C CH_3^+

D NH_3

30. In the homologous series of amines, increase in chain length from CH_3NH_2 to $C_4H_9NH_2$ is accompanied by

	Volatility	Solubility in water
A	increased	increased
B	decreased	decreased
C	increased	decreased
D	decreased	increased

31. The end-on overlap of two atomic orbitals lying along the axis of a bond is known as

A hybridisation

B a sigma bond

C a pi bond

D a double bond.

32. Alkenes react with ozone (O_3) to form ozonides which can be hydrolysed to give carbonyl compounds.

$$\diagdown C = C \diagup \xrightarrow{O_3} \text{OZONIDE} \xrightarrow{\text{hydrolysis}} \diagdown C = O + O = C \diagup$$

Which of the following alkenes will produce a mixture of propanone and ethanal when acted upon in this way?

A $CH_3CH = CHCH_2CH_3$

B $CH_3CH = CHCH_3$

C $(CH_3)_2C = CH_2$

D $CH_3CH = C(CH_3)_2$

33. Which of the following is the formula for a tertiary halogenoalkane?

A $CHBr_3$

B $(CH_3)_3CBr$

C $(CH_2Br)_3CH$

D $BrCH_2C(CH_3)_3$

34. A compound, X, has the formula C_6H_{12}. X **must** be

A   a hydrocarbon

B   an alkene

C   a cycloalkane

D   hexene.

35. A substance, X, is **readily** oxidised by acidified potassium dichromate solution to give a product which does **not** react with sodium carbonate, **nor** with Tollens' reagent.

Which of the following could represent the structure of X?

A $CH_3CH_2CH_2OH$

B $CH_3CH(OH)CH_3$

C CH_3CCH_3
   $\quad\ \underset{O}{\overset{\|}{}}$

D $CH_3CH_2C\overset{\diagup O}{\underset{\diagdown OH}{}}$

36. Which of the following compounds would liberate one mole of hydrogen gas if one mole of it reacts with two moles of sodium?

A C_2H_5OH

B $HOCH_2CH_2OH$

C CH_3COOH

D CH_3CHO

37. Which of the following will **not** form a derivative with 2,4-dinitrophenylhydrazine?

A $CH_3CH_2C\overset{\diagup\!\!\diagup O}{\underset{\diagdown H}{}}$

B $CH_3CH_2C\overset{\diagup\!\!\diagup O}{\underset{\diagdown OH}{}}$

C $CH_3CH_2CCH_3$
   $\qquad\ \underset{}{\overset{O}{\overset{\|}{}}}$

D   phenyl$-C\overset{\diagup\!\!\diagup O}{\underset{\diagdown H}{}}$

[Turn over

38. Spectral studies of an organic compound indicated the presence of a di-substituted benzene ring, two methyl groups and a molecular weight of 134.

Which of the following is a possible structure for the compound?

A

           CH$_3$

           COCH$_3$

B

           CH$_3$

           CHO

           CH$_3$

C

           CH$_3$CH$_2$

           COCH$_3$

D

           CH$_3$

           CH$_2$CHO

39. Which of the following compounds has a geometric isomer?

A

$$\begin{array}{ccc} & H & Cl \\ & | & | \\ H- & C-C & -H \\ & | & | \\ & Cl & H \end{array}$$

B

$$\begin{array}{cc} H & Cl \\ \diagdown \quad \diagup \\ C=C \\ \diagup \quad \diagdown \\ H & H \end{array}$$

C

$$\begin{array}{cc} H & Cl \\ \diagdown \quad \diagup \\ C=C \\ \diagup \quad \diagdown \\ H & Cl \end{array}$$

D

$$\begin{array}{cc} H & Cl \\ \diagdown \quad \diagup \\ C=C \\ \diagup \quad \diagdown \\ Cl & H \end{array}$$

40. A simplified mass spectrum of an organic compound is shown.

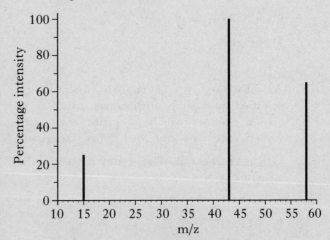

Which of the following compounds produces this spectrum?

A   Propane

B   Propan-1-ol

C   Propan-2-ol

D   Propanone

[END OF SECTION A]

Candidates are reminded that the answer sheet for Section A MUST be placed INSIDE the front cover of your answer book.

SECTION B

60 marks are available in this section of the paper.

All answers must be written clearly and legibly in ink.

Marks

1. Barium carbonate decomposes on heating.

$$BaCO_3(s) \longrightarrow BaO(s) + CO_2(g) \qquad \Delta H^\circ = +266 \, kJ \, mol^{-1}$$

   (a) Using the data from the table below, calculate the standard entropy change, ΔS°, in $J \, K^{-1} \, mol^{-1}$, for the reaction.

Substance	Standard entropy, $S^\circ / J \, K^{-1} \, mol^{-1}$
$BaCO_3(s)$	112·0
$BaO(s)$	72·1
$CO_2(g)$	213·8

   1

   (b) Calculate the temperature at which the decomposition of barium carbonate just becomes feasible.

   3

   (4)

2. Consider the following reaction.

$$CS_2(g) + 4H_2(g) \rightleftharpoons CH_4(g) + 2H_2S(g)$$

   At 900 °C the equilibrium concentrations are:

   $$[CS_2] = 0·012 \, mol \, l^{-1} \qquad [H_2] = 0·0020 \, mol \, l^{-1}$$
   $$[H_2S] = 0·00010 \, mol \, l^{-1} \qquad [CH_4] = 0·0054 \, mol \, l^{-1}$$

   (a) Write down the expression for the equilibrium constant, K, for this reaction.

   1

   (b) Calculate the value of the equilibrium constant, K, at 900 °C.

   1

   (2)

3. Fizzy drinks contain carbon dioxide dissolved in water which dissociates, as shown, to produce carbonic acid.

$$H_2O(\ell) + CO_2(aq) \rightleftharpoons H^+(aq) + HCO_3^-(aq) \qquad pK_a = 6·4$$

   (a) What is the Bronsted-Lowry definition of an acid?

   1

   (b) Write the formula for the conjugate base in this reaction.

   1

   (c) Calculate the pH of a $0·1 \, mol \, l^{-1}$ solution of carbonic acid.

   2

   (4)

   [Turn over

Marks

4. The following table of results was obtained for the reaction below.

$$H_2O_2(aq) \;+\; 2HI(aq) \;\longrightarrow\; 2H_2O(\ell) \;+\; I_2(aq)$$

Experiment	$[H_2O_2]/mol\,l^{-1}$	$[HI]/mol\,l^{-1}$	Initial rate/$mol\,l^{-1}\,s^{-1}$
1	$3 \cdot 2 \times 10^{-4}$	$4 \cdot 1 \times 10^{-4}$	$4 \cdot 3 \times 10^{-9}$
2	$6 \cdot 4 \times 10^{-4}$	$4 \cdot 1 \times 10^{-4}$	$8 \cdot 6 \times 10^{-9}$
3	$3 \cdot 2 \times 10^{-4}$	$8 \cdot 2 \times 10^{-4}$	$8 \cdot 6 \times 10^{-9}$
4	$6 \cdot 4 \times 10^{-4}$	$8 \cdot 2 \times 10^{-4}$	$1 \cdot 72 \times 10^{-8}$

(a) Determine the order of this reaction with respect to

   (i) H_2O_2

   (ii)   HI.   **1**

(b) Write the rate equation for the reaction.   **1**

(c) Calculate a value for the rate constant, k, including the appropriate units.   **2**

      (4)

5. A student set up the following electrochemical cell at 25 °C.

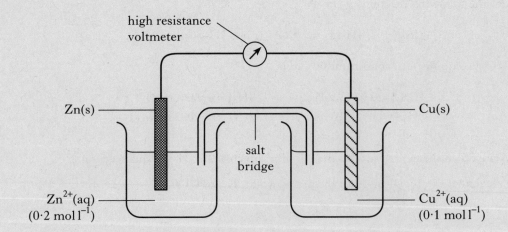

(a) Calculate ΔG, the actual free energy change, in $kJ\,mol^{-1}$, for this cell when the operating emf is 0·94 V.   **3**

(b) What changes should the student make to this cell for it to be operating under standard conditions?   **1**

(c) Calculate the emf of a $Zn(s)\,|\,Zn^{2+}(aq)\,||\,Cu^{2+}(aq)\,|\,Cu(s)$ cell operating under standard conditions.   **1**

      (5)

Marks

6. In a PPA, 3·43 g of hydrated nickel(II) sulphate, $NiSO_4.6H_2O$, was dissolved in water and made up to $100\,cm^3$ in a standard flask.

   $20·0\,cm^3$ of this solution was titrated against a $0·101\,mol\,l^{-1}$ solution of EDTA using murexide as an indicator. The results are shown below.

	Rough titre	1st titre	2nd titre	3rd titre
Initial burette reading/cm^3	0·0	0·0	24·6	0·0
Final burette reading/cm^3	24·8	24·6	48·8	24·3
Volume of EDTA added/cm^3	24·8	24·6	24·2	24·3

   (a) Give a reason why murexide is a suitable indicator in this titration reaction.   **1**

   (b) Calculate the percentage of nickel present in the hydrated salt from these experimental results.   **3**

   (c) The theoretical yield for the percentage of nickel in the hydrated salt is 22·3%. Suggest a reason for the difference between the theoretical percentage of nickel and the answer calculated in part (b).   **1**

   (5)

7. The electronic configuration for cobalt, Co, in its ground state, is

   $$1s^2\,2s^2\,2p^6\,3s^2\,3p^6\,3d^7\,4s^2.$$

   (a) In terms of s, p and d orbitals, write down the electronic configurations of
       (i) Cu   **1**
       (ii) Co^{2+} **1**
   in their ground states.

   (b) (i) Write an equation which represents the second ionisation energy for copper.   **1**
       (ii) Explain why copper has a higher second ionisation energy than cobalt.   **1**

   (4)

[Turn over

Marks

8. The following table gives information about aluminium chloride and magnesium chloride.

	Aluminium chloride	Magnesium chloride
Action of heat	sublimes at 178 °C	melts at 714 °C
Relative formula mass in gaseous state	267	95
Action of water	reacts to give white fumes	dissolves

(a) Using the above information determine

   (i) the main type of bonding in aluminium chloride and the main type of bonding in magnesium chloride   **1**

   (ii) the chemical formula for aluminium chloride in the gaseous state.   **1**

(b) What are the white fumes given off when water is added to aluminium chloride?   **1**

(c) Aluminium chloride can be used as a catalyst in the reaction between benzene and 2-chloropropane. Draw a structure for an organic product formed.   **1**

(d) Many ionic compounds adopt the same lattice structure as either sodium chloride or caesium chloride.

Sodium chloride has coordination 6:6 whereas caesium chloride has coordination 8:8.

   (i) Why do these chlorides have different lattice structures?   **1**

   (ii) Which of the above structures is iron(II) oxide most likely to adopt?

     Explain your answer.   **1**

        (6)

9. A common dietary supplement taken by athletes and slimmers is called chromium picolinate $[Cr(pic)_3]$. The structure of the picolinate ion, pic, is

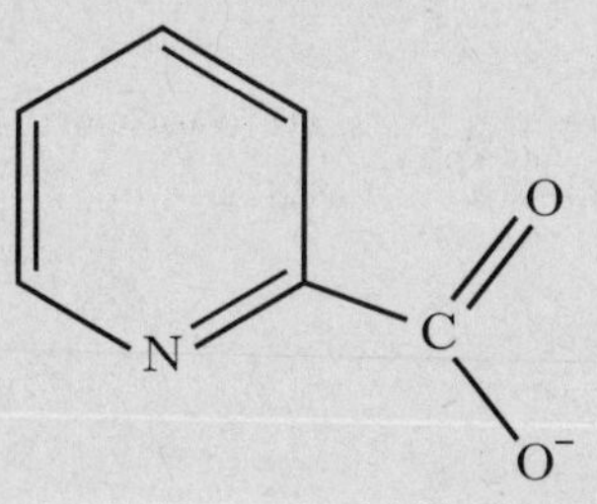

(a) What feature of the picolinate ion makes it suitable for use as a ligand?   **1**

(b) In the body, it is thought that the chromium in $[Cr(pic)_3]$ is changed into chromium(VI) by the action of hydrogen peroxide.

   (i) What is the oxidation state of chromium in $[Cr(pic)_3]$?   **1**

   (ii) What is the role of hydrogen peroxide in this reaction?   **1**

(c) A simpler complex of chromium is $[Cr(CN)_6]^{4-}$. What is its systematic name?   **1**

        (4)

Marks

10. Isoamyl acetate is found naturally as the flavour and scent of bananas.

    Its shortened structural formula is

    $$CH_3COOCH_2CH_2CH(CH_3)_2$$

    (a) (i) To which class of organic compounds does isoamyl acetate belong?   **1**

    (ii) Apart from flavouring agents, suggest another common use for this class of organic compound.   **1**

    Isoamyl acetate can be made in the laboratory by reacting ethanoic acid with another substance, **X**.

    (b) (i) Write the systematic name of substance **X**.   **1**

    (ii) Name the type of reaction which takes place.   **1**

       (4)

11. Propoxyphene is a pain-killing drug. Its structure is shown below.

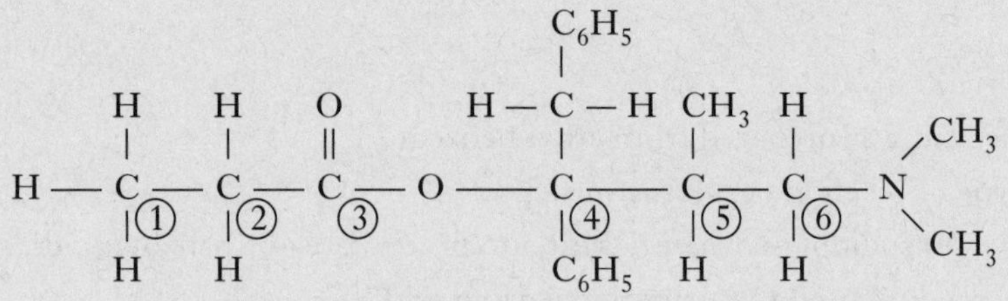

    (a) There are two chiral carbons in propoxyphene.

    Referring to the structure above, identify both chiral carbons.   **1**

    (b) Propoxyphene has a pharmacophore which binds to specific receptors.

    What is meant by the term **pharmacophore**?   **1**

    (c) Propoxyphene stimulates the body's own natural response to pain.

    What term is used to describe medicines which act in this way?   **1**

       (3)

    [Turn over

Marks

12. A student designed the following reaction scheme starting from ethyl benzoate.

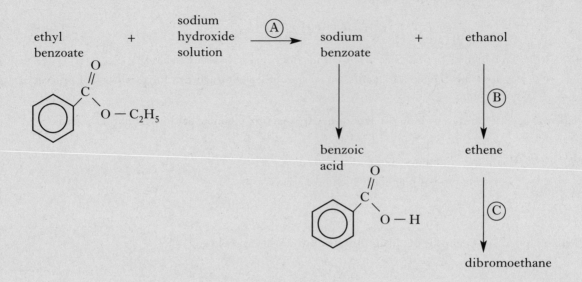

(a) In a PPA, benzoic acid was prepared from ethyl benzoate.

   (i) Name the type of reaction occurring in step (A).   **1**

   (ii) What is added to sodium benzoate solution to precipitate out benzoic acid?   **1**

   (iii) Name the procedure used to purify the benzoic acid.   **1**

   (iv) Assuming the percentage yield is 70%, what is the minimum mass of ethyl benzoate required to produce at least 4·0 g of benzoic acid?   **2**

(b)   (i) Name the type of reaction occurring in step (B).   **1**

   (ii) Using structural formulae, outline the mechanism for the reaction occurring in step (C).   **2**

                         (8)

Marks

13. A compound **X** containing only carbon, hydrogen and oxygen, was subjected to elemental analysis. Complete combustion of 1·76 g of **X** gave 3·52 g of carbon dioxide and 1·44 g of water. No other product was formed.

   (*a*)   (i)   Calculate the masses of carbon and hydrogen in the original sample and hence deduce the mass of oxygen present.   **2**

         (ii)   Show, by calculation, that the empirical formula of compound **X** is C_2H_4O.   **1**

   (*b*)   Given that the relative molecular mass of compound **X** is 88, deduce its molecular formula.   **1**

   (*c*)   A low-resolution proton nmr spectrum of compound **X** is shown.

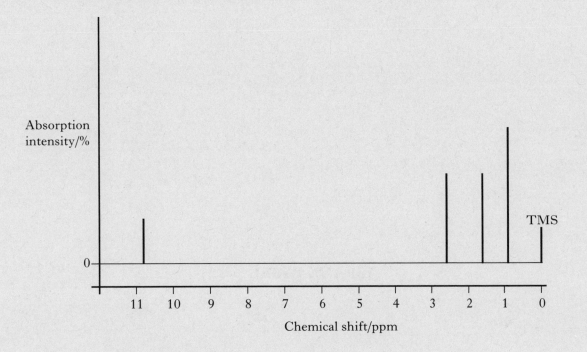

Analysis of this spectrum produced the data shown in the table below.

Chemical shift	Relative area under the peak
0·9	3
1·6	2
2·6	2
10·8	1

        (i)   Using information from the table and your answer to (*b*), draw a structural formula for compound **X** and give its systematic name.   **2**

       (ii)   What is the function of "TMS" (tetramethylsilane) in the proton nmr spectrum?   **1**

                                                                                        (7)

[END OF QUESTION PAPER]

[BLANK PAGE]

ADVANCED HIGHER

2008

[BLANK PAGE]

X012/701

NATIONAL QUALIFICATIONS 2008	FRIDAY, 30 MAY 9.00 AM – 11.30 AM	CHEMISTRY ADVANCED HIGHER

Reference may be made to the Chemistry Higher and Advanced Higher Data Booklet.

SECTION A – 40 marks

Instructions for completion of **SECTION A** are given on page two.

For this section of the examination you must use an **HB pencil**.

SECTION B – 60 marks

All questions should be attempted.

Answers must be written clearly and legibly in ink.

SECTION A

Read carefully

1 Check that the answer sheet provided is for **Chemistry Advanced Higher (Section A)**.

2 For this section of the examination you must use an **HB pencil** and, where necessary, an eraser.

3 Check that the answer sheet you have been given has **your name**, **date of birth**, **SCN** (Scottish Candidate Number) and **Centre Name** printed on it.

   Do not change any of these details.

4 If any of this information is wrong, tell the Invigilator immediately.

5 If this information is correct, **print** your name and seat number in the boxes provided.

6 The answer to each question is **either** A, B, C or D. Decide what your answer is, then, using your pencil, put a horizontal line in the space provided (see sample question below).

7 There is **only one correct** answer to each question.

8 Any rough working should be done on the question paper or the rough working sheet, **not** on your answer sheet.

9 At the end of the exam, put the **answer sheet for Section A inside the front cover of your answer book**.

Sample Question

To show that the ink in a ball-pen consists of a mixture of dyes, the method of separation would be

     A   chromatography

     B   fractional distillation

     C   fractional crystallisation

     D   filtration.

The correct answer is **A**—chromatography. The answer **A** has been clearly marked in **pencil** with a horizontal line (see below).

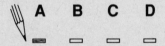

Changing an answer

If you decide to change your answer, carefully erase your first answer and using your pencil, fill in the answer you want. The answer below has been changed to **D**.

     A   B   C   D
     ▭   ▭   ▭   ▬

1. An atom has the electronic configuration

$$1s^2\, 2s^2\, 2p^6\, 3s^2\, 3p^1$$

What is the charge of the most likely ion formed from this atom?

A   −1

B   +1

C   +2

D   +3

2. The electronic configurations, **X** and **Y**, for two uncharged atoms of sodium are as follows.

X $1s^2\, 2s^2\, 2p^6\, 3s^1$
Y $1s^2\, 2s^2\, 2p^6\, 4s^1$

Which of the following statements is true?

A   **X** is an excited state.

B   Both **X** and **Y** have vacant 2d orbitals.

C   Energy is absorbed in changing **Y** to **X**.

D   Less energy is required to ionise **Y** compared to **X**.

3. A Lewis base may be regarded as a substance which is capable of donating an unshared pair of electrons to form a covalent bond.

Which of the following could act as a Lewis base?

A Co^{3+}

B PH_3

C BCl_3

D NH_4^+

4. Silicon can be converted into an n-type semiconductor by adding

A   boron

B   carbon

C   arsenic

D   aluminium.

5. Which of the following statements referring to the structures of sodium chloride and caesium chloride is correct?

A   There are eight chloride ions surrounding each sodium ion.

B   There are eight chloride ions surrounding each caesium ion.

C   The chloride ions are arranged tetrahedrally round the sodium ions.

D   The chloride ions are arranged tetrahedrally round the caesium ions.

6. The transition metal salts, MnF_2, FeF_2 and CoF_2, have identical crystal structures because the metal ions have

A   similar radii

B   similar colours

C   the same nuclear charge

D   the same number of d electrons.

7. Which of the following hydrides, when added to water, would give the most acidic solution?

A   Sodium hydride

B   Magnesium hydride

C   Silicon hydride

D   Sulphur hydride

8. Sodium hydride reacts with sodium sulphate as shown.

$$4NaH\ +\ Na_2SO_4\ \rightarrow\ 4NaOH\ +\ Na_2S$$

This reaction demonstrates sodium hydride acting as

A   a base

B   an acid

C   a reducing agent

D   an oxidising agent.

[Turn over

9. Three elements, **X**, **Y** and **Z**, are in the same period of the Periodic Table.

   The oxide of **X** is amphoteric, the oxide of **Y** is basic and the oxide of **Z** is acidic.

   Which of the following shows the elements arranged in order of increasing atomic number?

   A **Y, X, Z**

   B **Y, Z, X**

   C **Z, X, Y**

   D **X, Y, Z**

10. Which of the following involves oxidation?

    A $MnO_4^- \rightarrow MnO_4^{2-}$

    B $Ag^+ \rightarrow [Ag(NH_3)_2]^+$

    C $[Fe(CN)_6]^{4-} \rightarrow [Fe(CN)_6]^{3-}$

    D $[Ni(H_2O)_6]^{2+} \rightarrow [Ni(CN)_4]^{2-}$

11. The number of unpaired electrons in a gaseous Ni^{2+} ion is

    A 0

    B 2

    C 4

    D 6.

12. $P + Q \rightleftharpoons R + S$

    At 298 K the equilibrium constant for this reaction is $1 \cdot 2 \times 10^{10}$.

    Which of the following is true?

    A The value of ΔS° must be positive.

    B The value of ΔG° must be positive.

    C Adding a catalyst will change the equilibrium constant.

    D Increasing the concentration of P will not change the equilibrium constant.

13.

    $$CH_3COOH + C_2H_5OH \rightleftharpoons CH_3COOC_2H_5 + H_2O$$

    The above reaction can be said to have reached equilibrium when

    A the equilibrium constant K is equal to 1

    B the reaction between the acid and the alcohol has stopped

    C the concentrations of the products equal those of the reactants

    D the rate of production of ethyl ethanoate equals its rate of hydrolysis.

14. When sulphur dioxide and oxygen react the following equilibrium is established.

    $$2SO_2(g) + O_2(g) \rightleftharpoons 2SO_3(g)$$

    The equilibrium constant for the reaction is 3300 at 630 °C and 21 at 850 °C.

    Which line in the table is correct for the reaction?

	Sign of ΔH	Product yield as temperature increases
A	+	decreases
B	+	increases
C	−	decreases
D	−	increases

15. $500\,cm^3$ of $0 \cdot 022\ mol\,l^{-1}$ hydrochloric acid is mixed with $500\,cm^3$ of $0 \cdot 020\ mol\,l^{-1}$ sodium hydroxide solution. The pH of the resulting solution will be

    A 2

    B 3

    C 4

    D 5.

16. The Bronsted-Lowry definition of a base is a substance which acts as a

    A proton donor to form a conjugate acid

    B proton donor to form a conjugate base

    C proton acceptor to form a conjugate acid

    D proton acceptor to form a conjugate base.

17. The mean bond enthalpy of the N–H bond is equal to one third of the value of ΔH for which change of the following changes?

A $NH_3(g) \rightarrow N(g) + 3H(g)$

B $2NH_3(g) \rightarrow N_2(g) + 3H_2(g)$

C $NH_3(g) \rightarrow \frac{1}{2}N_2(g) + 1\frac{1}{2}H_2(g)$

D $2NH_3(g) + 1\frac{1}{2}O_2(g) \rightarrow N_2(g) + 3H_2O(g)$

18. The entropy of a perfect crystal is zero at

A 0 K

B 25 K

C 273 K

D 298 K.

19. Which of the following reactions results in a **decrease** in entropy?

A $N_2O_4(g) \rightarrow 2NO_2(g)$

B $N_2(g) + 3H_2(g) \rightarrow 2NH_3(g)$

C $CaCO_3(s) \rightarrow CaO(s) + CO_2(g)$

D $C(s) + H_2O(g) \rightarrow CO(g) + H_2(g)$

20. Which of the following is **not** a required condition for measuring standard electrode potentials?

A Volume of 1 litre

B Temperature of 298 K

C Concentration of $1\ mol\,l^{-1}$

D Pressure of 1 atmosphere

21.

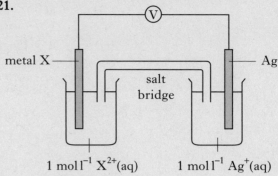

$1\ mol\,l^{-1}\ X^{2+}(aq)$ $1\ mol\,l^{-1}\ Ag^+(aq)$

The E° values are

$X^{2+}(aq) + 2e^- \rightarrow X(s)$ $E° = -0.23\,V$

$Ag^+(aq) + e^- \rightarrow Ag(s)$ $E° = 0.80\,V$

In the above cell, which of the following is reduced?

A $X(s)$

B $Ag(s)$

C $X^{2+}(aq)$

D $Ag^+(aq)$

22. Under standard conditions, the emf of the cell

$$Al(s)\,|\,Al^{3+}(aq)\,||\,Cu^{2+}(aq)\,|\,Cu(s)$$

would be

A 1·34 V

B 2·02 V

C 2·34 V

D 4·38 V.

23. For a cell in which the following reaction occurs

$$X(s) + 2Y^+(aq) \rightarrow X^{2+}(aq) + 2Y(s)$$

the E° value is 1·5 V.

$\Delta G°$ for the reaction, per mole of X, is

A $-289.5\,kJ\,mol^{-1}$

B $-144.8\,kJ\,mol^{-1}$

C $+144.8\,kJ\,mol^{-1}$

D $+289.5\,kJ\,mol^{-1}$.

[Turn over

24.

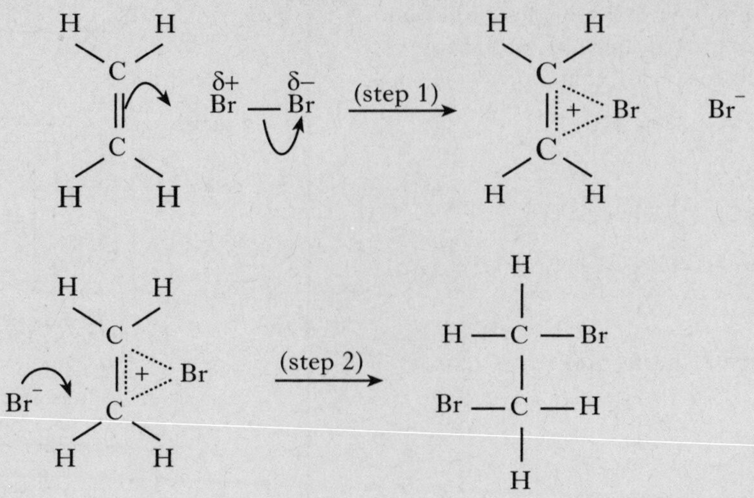

The two steps in the reaction mechanism shown can be described as

A   ethene acting as a nucleophile and Br^- acting as a nucleophile

B   ethene acting as a nucleophile and Br^- acting as an electrophile

C   ethene acting as an electrophile and Br^- acting as a nucleophile

D   ethene acting as an electrophile and Br^- acting as an electrophile.

25. In the homologous series of alkanols, increase in chain length from CH_3OH to $C_{10}H_{21}OH$ is accompanied by

A   increased volatility and increased solubility in water

B   increased volatility and decreased solubility in water

C   decreased volatility and decreased solubility in water

D   decreased volatility and increased solubility in water.

26. Which of the following is **not** caused by hydrogen bonding?

A   The low density of ice compared to water

B   The solubility of methoxymethane in water

C   The higher boiling point of methanol compared to ethane

D   The higher melting point of hydrogen compared to helium

27. A compound C_3H_8O does **not** react with sodium and is **not** reduced by lithium aluminium hydride. It is likely to be an

A   acid

B   ether

C   alcohol

D   aldehyde.

28. Which of the following is least acidic?

A CH_3OH

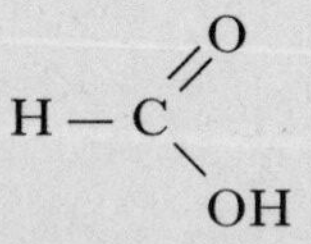

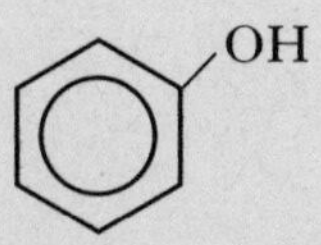

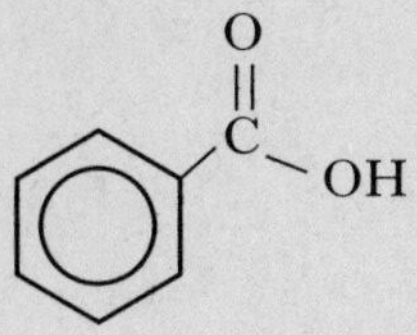

29. Which statement about ethanol and its isomeric ether is true?

They

A have similar volatilities

B have similar infra-red spectra

C form the same products when burned in excess oxygen

D form the same products when reacted with acidified dichromate.

30. When 2-bromobutane is reacted with potassium cyanide and the compound formed is hydrolysed with dilute acid, the final product is

A butanoic acid

B pentanoic acid

C 2-methylbutanoic acid

D 2-methylpropanoic acid.

31. Which of the following compounds would liberate one mole of hydrogen gas if one mole of it reacts with excess sodium?

A C_2H_5OH

B CH_3CHO

C CH_3COOH

D $HOCH_2CH_2OH$

32. Two isomeric esters, **X** and **Y**, have the molecular formula $C_4H_8O_2$. Ester **X** on hydrolysis with sodium hydroxide solution gives CH_3CH_2COONa, and ester **Y** on similar treatment gives CH_3CH_2OH.

Which line in the table shows the correct names of **X** and **Y**?

	X	Y
A	propyl methanoate	ethyl ethanoate
B	methyl propanoate	ethyl ethanoate
C	methyl propanoate	ethyl methanoate
D	propyl methanoate	methyl propanoate

33. A white crystalline compound, soluble in water, was found to react with both dilute hydrochloric acid and sodium hydroxide solution.

Which of the following might it have been?

A C_6H_5OH

B $C_6H_5NH_2$

C C_6H_5COOH

D H_2NCH_2COOH

34. Which of the following amines has the lowest boiling point?

A $C_4H_9NH_2$

B $C_3H_7NHCH_3$

C $C_2H_5NHC_2H_5$

D $C_2H_5N(CH_3)_2$

35. Spectral studies of an organic compound indicated the presence of a di-substituted benzene ring, two methyl groups and a molecular weight of 134.

Which of the following is a possible structure for the compound?

A

B

C

D

[Turn over

36. Which of the following molecules does **not** exhibit optical isomerism?

A — Br, Br

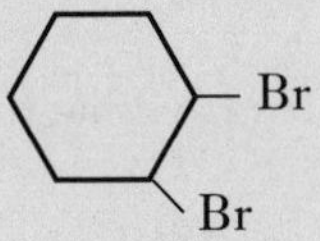

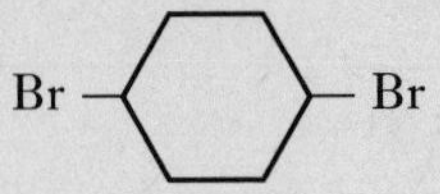

B Br — — Br

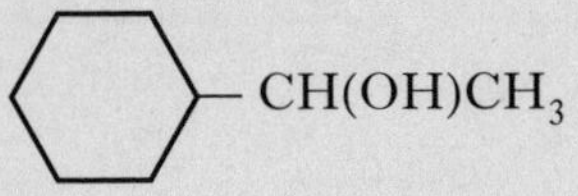

C — $CH(OH)CH_3$

D — CHBrCHO

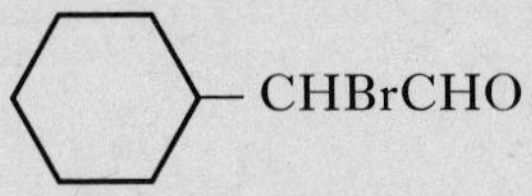

37. Which of the following ~~could~~ can ~~not~~ exist in isomeric forms?

A C_2F_4

B C_3H_6

C C_3H_7Br

D $C_2H_4Cl_2$

38. Elemental analysis of an organic compound showed it contained 70·6% carbon, 23·5% oxygen and 5·9% hydrogen by mass.

The structural formula of the compound could be

A CH_2CHCH_2COOH

B — OH, OH

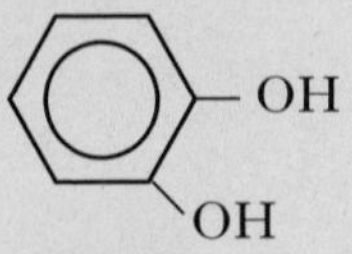

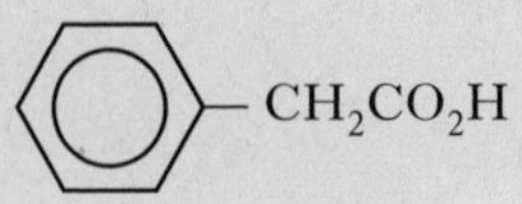

C — CH_2CO_2H

D — CH_2CHO

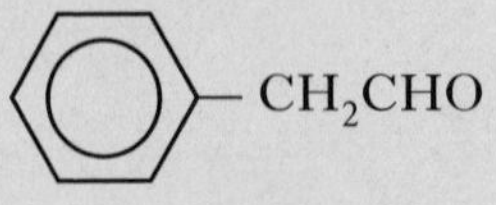

39. An organic compound with empirical formula, C_2H_4O, has major peaks at $1715\,cm^{-1}$ and $3300\,cm^{-1}$ in its infrared spectrum.

The structural formula of the compound could be

A CH_3CHO

B CH_3COOH

C $CH_3COOCH_2CH_3$

D $CH_3CH_2CH_2COOH.$

40. A drug containing a carboxyl group can bind to an amino group on a receptor site in three different ways.

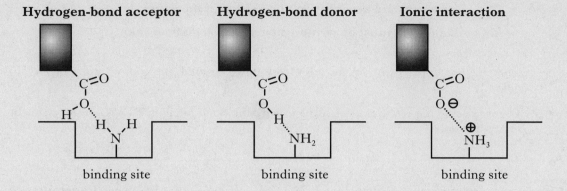

The drug with the following structure

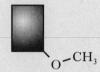

could bind to the same site

A   only by ionic interaction

B   only as a hydrogen-bond donor

C   only as a hydrogen-bond acceptor

D   both as a hydrogen-bond donor and acceptor.

[*END OF SECTION A*]

Candidates are reminded that the answer sheet for Section A MUST be placed INSIDE the front cover of your answer book.

[**Turn over for SECTION B on *Page ten***

SECTION B

60 marks are available in this section of the paper.

All answers must be written clearly and legibly in ink.

Marks

1. (a) What is the arrangement of the **electron pairs** around the iodine atom in an IF_5 molecule?   *octahedral*   **1**

    (b) By considering the electron pairs, explain why the bond angle in BF_3 is greater than the bond angle in NF_3.   **1**

    (2)

2. An aqueous solution of the compound $[CoCl_2(NH_3)_4]Cl$ gave the following **transmittance** spectrum.

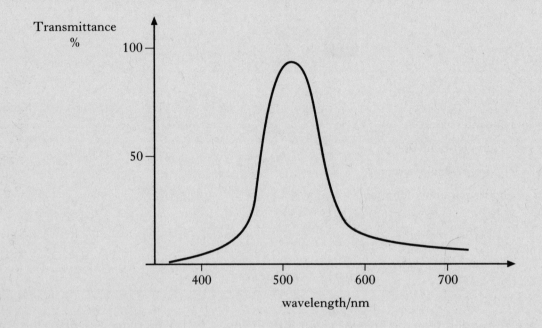

    (a) From the above spectrum, deduce the colour of the solution.   **1**

    (b) The solution contains the complex ion $[CoCl_2(NH_3)_4]^+$.

        (i)   What is the oxidation number of cobalt in this complex ion?   **1**

        (ii) Name this complex ion.   **1**

        (iii) Write down the electronic configuration of cobalt in this complex ion in terms of s, p and d orbitals.   **1**

    (4)

3. Some metal salts emit light when heated in a Bunsen flame.

    Lithium nitrate changes the flame colour to crimson.

    Magnesium nitrate has no effect on the flame colour.

    (a) Explain, in terms of electrons, why some metal salts emit light when heated in a Bunsen flame.   **1**

    (b) Suggest why magnesium nitrate has no effect on the flame colour.   **1**

    (c) Calculate the energy, in $kJ\,mol^{-1}$, associated with crimson light of wavelength 671 nm.   **2**

    (4)

Marks

4. *cis*-Platin is a highly successful anti-cancer drug. The formula for *cis*-platin is $[Pt(NH_3)_2Cl_2]$.

   (a) *cis*-Platin works by forming a complex with parts of a DNA molecule. These parts of the DNA form bonds through nitrogen atoms to $Pt(NH_3)_2$ as shown below.

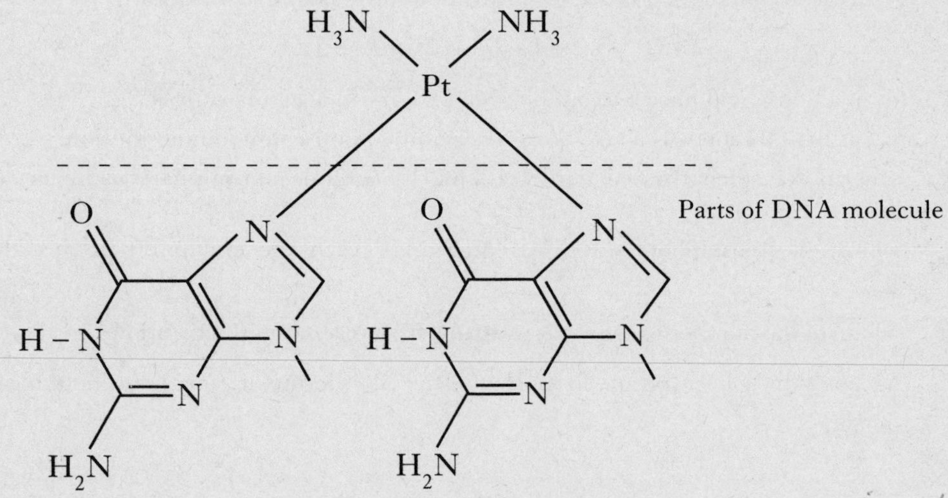

Parts of DNA molecule

   (i)   Explain why DNA can be classified as a bidentate ligand in this complex. **1**

   (ii) What feature of the DNA makes it suitable as a ligand? **1**

   (b) Draw a possible structure for the geometric isomer of *cis*-platin. **1**

   (3)

5. The equation for the decomposition of ammonium dichromate is

$$(NH_4)_2Cr_2O_7(s) \longrightarrow N_2(g) + 4H_2O(\ell) + Cr_2O_3(s)$$

Consider the following data for the reaction at 298 K.

Substance	$\Delta H_f^\circ/kJ\,mol^{-1}$	$S^\circ/J\,K^{-1}mol^{-1}$
$(NH_4)_2Cr_2O_7(s)$	−1806	336
$N_2(g)$	0	192
$H_2O(\ell)$	−286	70
$Cr_2O_3(s)$	−1140	81

   (a) For the decomposition of $(NH_4)_2Cr_2O_7$, calculate

   (i) ΔH° **1**

   (ii) ΔS° **1**

   (iii) ΔG°. **2**

   (b) Chromium burns in excess oxygen to form chromium(III) oxide. From the information in the table, deduce a value for the enthalpy of combustion of chromium. **1**

   (5)

[Turn over

Marks

6. Sodium hypochlorite, NaClO, is the active ingredient in household bleach. The concentration of the hypochlorite ion, ClO^-, can be determined in two stages.

   In stage 1, an acidified iodide solution is added to a solution of the bleach and iodine is formed.

   $$ClO^-(aq) + 2I^-(aq) + 2H^+(aq) \rightarrow I_2(aq) + Cl^-(aq) + H_2O(\ell)$$

   In stage 2, the iodine formed is titrated with sodium thiosulphate solution.

   $$2S_2O_3^{2-}(aq) + I_2(aq) \rightarrow 2I^-(aq) + S_4O_6^{2-}(aq)$$

   $10.0\,cm^3$ of a household bleach was diluted to $250\,cm^3$ in a standard flask.

   $25.0\,cm^3$ of this solution was added to excess acidified potassium iodide solution.

   The solution was then titrated with $0.10\,mol\,l^{-1}$ sodium thiosulphate using an appropriate indicator.

   The volume of thiosulphate solution required to reach the end point of the titration was $20.5\,cm^3$.

   (a) Calculate the number of moles of **iodine** which reacted in the titration.   1

   (b) Calculate the concentration, in $mol\,l^{-1}$, of the ClO^- in the original household bleach.   2

                                                                                                    (3)

7. Consider the Born-Haber cycle below which represents the formation of caesium fluoride.

   $$Cs(s) \quad + \quad \tfrac{1}{2}F_2(g) \longrightarrow Cs^+F^-(s)$$

   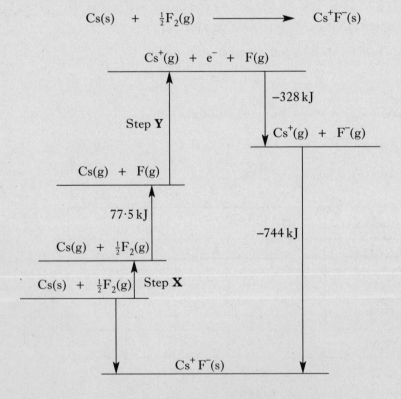

   (a) Use the Data Booklet to find the enthalpy values for Step **X** and Step **Y**.   2

   (b) Name the enthalpy change that has the value $-744\,kJ$ in this cycle.   1

   (c) Use this Born-Haber cycle to calculate the enthalpy of formation of caesium fluoride in $kJ\,mol^{-1}$.   1

                                                                                                    (4)

Marks

8. In a PPA, a student added $50\,cm^3$ of an aqueous iodine solution to $50\,cm^3$ of cyclohexane in a separating funnel. After shaking thoroughly, the funnel was left until the following equilibrium was established.

$$I_2(\text{aqueous}) \quad \rightleftharpoons \quad I_2(\text{cyclohexane})$$

Two layers were formed, each containing dissolved iodine. $10\cdot0\,cm^3$ of each layer was titrated with sodium thiosulphate solution until the end point was reached.

The cyclohexane layer required $18\cdot8\,cm^3$ of $0\cdot025\,mol\,l^{-1}$ sodium thiosulphate.

The aqueous layer required $10\cdot5\,cm^3$ of $0\cdot050\,mol\,l^{-1}$ sodium thiosulphate.

(a) Which indicator is used to show that the end point has been reached?   1

(b) Calculate the concentration of iodine in

   (i) the cyclohexane layer   1

   (ii) the aqueous layer.   1

(c) Calculate the partition coefficient for iodine between the two solvents.   1

(d) If $100\,cm^3$ of cyclohexane had been used instead of the $50\,cm^3$, what effect would this have on

   (i) the concentration of iodine in the aqueous layer   1

   (ii) the value of the partition coefficient?   1

   (6)

[Turn over

Marks

9. Hydrofluoric acid, HF, is a weak acid.

$$HF(aq) \quad + \quad H_2O(\ell) \quad \rightleftharpoons \quad H_3O^+(aq) \quad + \quad F^-(aq)$$

A student neutralised 25 cm^3 of hydrofluoric acid solution with sodium hydroxide solution and followed the reaction by measuring the pH.

The graph obtained for this reaction is shown below.

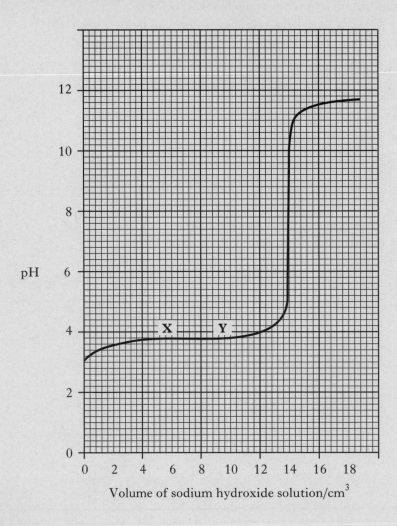

Volume of sodium hydroxide solution/cm^3

(a) Write the expression for the dissociation constant, K_a, of hydrofluoric acid.   **1**

(b) When exactly half the acid has been neutralised, $pK_a = pH$.

   Using only information from the graph, deduce pK_a and thus calculate K_a for hydrofluoric acid.   **2**

(c) The region **XY** on the graph is sometimes referred to as the buffer region.

   Apart from HF, what else is present in the solution which enables it to act as a buffer?   **1**

(d)

Indicator	pK_{In}
Methyl orange	3·7
Alizarin red	6·6
Cresol red	8·0
Alizarin yellow	11·1

Which of the above indicators could be used to detect the end point of this neutralisation reaction?   **1**

(5)

Marks

10. A mixture of butan-1-ol and butan-2-ol can be synthesised from 1-bromobutane in a two stage process.

$$CH_3CH_2CH_2CH_2Br \xrightarrow[\textbf{Stage 1}]{KOH/C_2H_5OH} CH_3CH_2CH{=}CH_2 \xrightarrow[\textbf{Stage 2}]{H_2O/H^+} \begin{array}{c} \text{butan-1-ol} \\ + \\ \text{butan-2-ol} \end{array}$$

(a) What type of reaction is taking place in **Stage 1**?   1

(b) The bonding in but-1-ene can be described in terms of sp^2 and sp^3 hybridisation and sigma and pi bonds.

  (i) What is meant by sp^2 hybridisation?   1

  (ii) What is the difference in the way atomic orbitals overlap to form sigma and pi bonds?   1

(c) Draw a structural formula for the major product of **Stage 2**.   1

(d) 1-Bromobutane reacts with hydroxide ions in a nucleophilic substitution reaction to produce butan-1-ol. The following results were obtained for this reaction.

Experiment	[1-Bromobutane]/mol l^{-1}	[OH$^-$]/mol l^{-1}	Initial rate/mol l^{-1} s^{-1}
1	0·25	0·10	$3·3 \times 10^{-6}$
2	0·50	0·10	$6·6 \times 10^{-6}$
3	0·50	0·20	$1·3 \times 10^{-5}$

  (i) What is the overall order of this reaction?   1

  (ii) Calculate a value for the rate constant of this reaction, giving the appropriate units.   2

  (iii) Outline the mechanism for this nucleophilic substitution reaction using structural formulae.   2

  (9)

11. A student devised the following reaction sequence starting from propan-1-ol, C_3H_7OH.

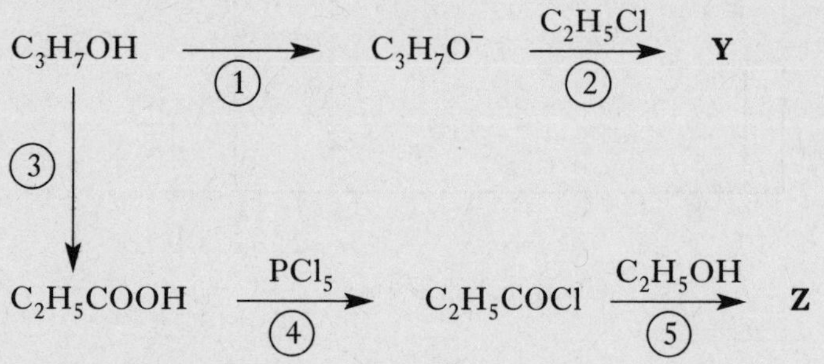

$$C_3H_7OH \xrightarrow{\;①\;} C_3H_7O^- \xrightarrow[②]{C_2H_5Cl} Y$$

$$\downarrow ③$$

$$C_2H_5COOH \xrightarrow[④]{PCl_5} C_2H_5COCl \xrightarrow[⑤]{C_2H_5OH} Z$$

(a) Name a suitable reagent to carry out

  (i) Step ①

  (ii) Step ③.   2

(b) Name **Y**.   1

(c) Draw a structural formula for **Z**.   1

  (4)

[Turn over

Marks

12. In a PPA, propanone reacts with 2,4-dinitrophenylhydrazine to make the 2,4-dinitrophenylhydrazone derivative as shown below.

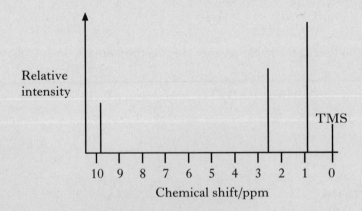

propanone   2,4-dinitrophenylhydrazine   2,4-dinitrophenylhydrazone derivative

(a) What type of reaction is this?   1

(b) The 2,4-dinitrophenylhydrazone derivative formed in the reaction is impure.

   (i) How would the derivative be purified?   1

   (ii) How can the technique of derivative formation be used to identify an unknown ketone?   1

(c) Propanone has an isomer. The shortened structural formula of this isomer is CH_3CH_2CHO.

   (i) Which chemical reagent could be used to distinguish between propanone and this isomer and what would be the result?   1

   (ii) Nuclear magnetic resonance spectroscopy can also be used to distinguish between these two isomers. The proton nmr spectrum for CH_3CH_2CHO is shown.

Relative intensity

TMS

10 9 8 7 6 5 4 3 2 1 0

Chemical shift/ppm

Sketch the proton nmr spectrum you would obtain for propanone.   1

Marks

12. **(c)** **(continued)**

(iii)   A simplified mass spectrum for propanone is shown below.

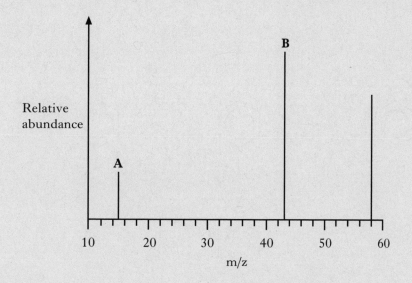

Identify the ion fragments responsible for peaks **A** and **B**.   **1**

                                                                              (6)

[Turn over for Question 13 on *Page eighteen*

Marks

13. A student devised the following reaction scheme starting with benzene.

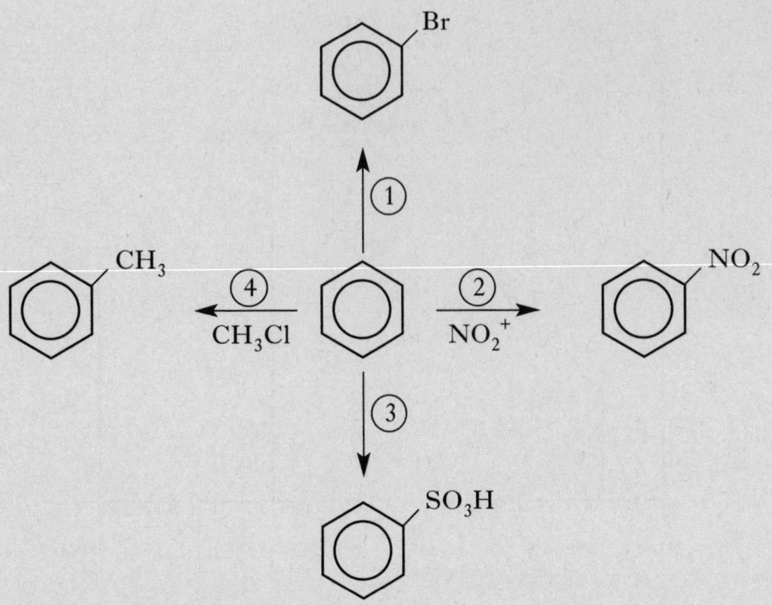

(a) What type of reaction does benzene undergo in reactions ① – ④ ?   **1**

(b) Name a suitable reagent and catalyst for reaction ① .   **1**

(c) Reaction ② involves nitration of benzene.
Which reagents are used to produce the NO_2^+ ion?   **1**

(d) What is the molecular formula for the product of reaction ③ ?   **1**

(e) The product of reaction ④ was reacted with bromine in the presence of light.
Draw a structural formula for an organic product of this reaction.   **1**

(5)

[END OF QUESTION PAPER]

ADVANCED HIGHER

2009

[BLANK PAGE]

X012/701

NATIONAL QUALIFICATIONS 2009	WEDNESDAY, 3 JUNE 9.00 AM – 11.30 AM	CHEMISTRY ADVANCED HIGHER

Reference may be made to the Chemistry Higher and Advanced Higher Data Booklet .

SECTION A – 40 marks

Instructions for completion of **SECTION A** are given on page two.

For this section of the examination you must use an **HB pencil**.

SECTION B – 60 marks

All questions should be attempted.

Answers must be written clearly and legibly in ink.

SECTION A

Read carefully

1 Check that the answer sheet provided is for **Chemistry Advanced Higher (Section A)**.

2 For this section of the examination you must use an **HB pencil** and, where necessary, an eraser.

3 Check that the answer sheet you have been given has **your name**, **date of birth**, **SCN** (Scottish Candidate Number) and **Centre Name** printed on it.

Do not change any of these details.

4 If any of this information is wrong, tell the Invigilator immediately.

5 If this information is correct, **print** your name and seat number in the boxes provided.

6 The answer to each question is **either** A, B, C or D. Decide what your answer is, then, using your pencil, put a horizontal line in the space provided (see sample question below).

7 There is **only one correct** answer to each question.

8 Any rough working should be done on the question paper or the rough working sheet, **not** on your answer sheet.

9 At the end of the exam, put the **answer sheet for Section A inside the front cover of your answer book**.

Sample Question

To show that the ink in a ball-pen consists of a mixture of dyes, the method of separation would be

    A chromatography

    B fractional distillation

    C fractional crystallisation

    D filtration.

The correct answer is **A**—chromatography. The answer **A** has been clearly marked in **pencil** with a horizontal line (see below).

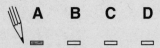

Changing an answer

If you decide to change your answer, carefully erase your first answer and using your pencil, fill in the answer you want. The answer below has been changed to **D**.

    A   B   C   D

1. The diagram shows one of the series of lines in the hydrogen emission spectrum.

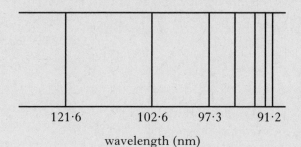

wavelength (nm)

Each line

A represents an energy level within a hydrogen atom

B results from an electron moving to a higher energy level

C lies within the visible part of the electromagnetic spectrum

D results from an excited electron dropping to a lower energy level.

2. Which of the following compounds shows most covalent character?

A CH_4

B NaH

C NH_3

D PH_3

3. In which of the following species is a dative covalent bond present?

A H_3O^+

B H_2O

C OH^-

D O_2

4. Which of the following diagrams best represents the arrangement of electron pairs around the central iodine atom in the I_3^- ion?

A

B

C

D

5. When a voltage is applied to an n-type semiconductor, which of the following migrate through the lattice?

A Electrons

B Negative ions

C Positive holes

D Both electrons and positive holes

6. Which of the following compounds would produce fumes of hydrogen chloride when added to water?

A LiCl

B $MgCl_2$

C PCl_3

D CCl_4

[Turn over

7. Zinc oxide reacts as shown.

$$ZnO(s) + 2HCl(aq) \rightarrow ZnCl_2(aq) + H_2O(\ell)$$

$$ZnO(s) + 2NaOH(aq) + H_2O(\ell) \rightarrow Na_2Zn(OH)_4(aq)$$

This shows that zinc oxide is

A basic

B acidic

C neutral

D amphoteric.

8. The correct formula for the tetraamminedichlorocopper(II) complex is

A $[Cu(NH_3)_4Cl_2]^{2-}$

B $[Cu(NH_3)_4Cl_2]$

C $[Cu(NH_3)_4Cl_2]^{2+}$

D $[Cu(NH_3)_4Cl_2]^{4+}$.

9. Which of the following aqueous solutions contains the **greatest** number of **negatively** charged ions?

A $500\,cm^3$ $0.10\,mol\,l^{-1}$ $Na_2SO_4(aq)$

B $250\,cm^3$ $0.12\,mol\,l^{-1}$ $BaCl_2(aq)$

C $300\,cm^3$ $0.15\,mol\,l^{-1}$ $KI(aq)$

D $400\,cm^3$ $0.10\,mol\,l^{-1}$ $Zn(NO_3)_2(aq)$

10. When one mole of phosphorus pentachloride was heated to 523 K in a closed vessel, 50% dissociated as shown.

$$PCl_5(g) \rightleftharpoons PCl_3(g) + Cl_2(g)$$

How many moles of gas were present in the equilibrium mixture?

A 0.5

B 1.0

C 1.5

D 2.0

11. Which of the following graphs shows the temperature change as 2 $mol\,l^{-1}$ sodium hydroxide is added to 25 cm^3 of 2 $mol\,l^{-1}$ hydrochloric acid?

A
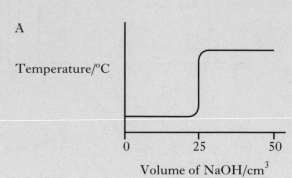

B
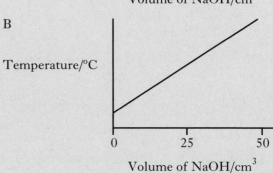

C
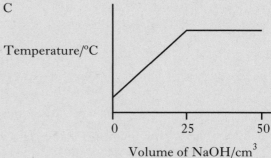

D
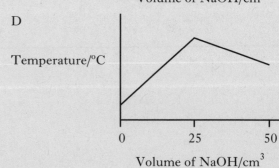

12. In the equilibrium $N_2O_4(g) \rightleftharpoons 2NO_2(g)$ the forward reaction is endothermic.

Which one of the following causes an increase in the value of the equilibrium constant?

A The removal of NO_2

B An increase of pressure

C A decrease of temperature

D An increase of temperature

13. In which of the following separation techniques is partition between two separate phases **not** a part of the process?

A Recrystallisation of benzoic acid from hot water

B Separation of alkanes using gas-liquid chromatography

C Separation of plant dyes using paper chromatography

D Solvent extraction of caffeine from an aqueous solution using dichloromethane

14. An aqueous solution of an organic acid, X, was shaken with chloroform until the following equilibrium was established.

$$X \text{ (water)} \rightleftharpoons X \text{ (chloroform)}$$

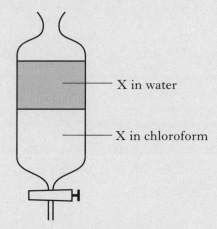

25.0 cm^3 of the upper layer needed 20.0 cm^3 of 0.050 mol l^{-1} NaOH(aq) for neutralisation. 25.0 cm^3 of the lower layer needed 13.3 cm^3 of 0.050 mol l^{-1} NaOH(aq) for neutralisation.

The value of the partition coefficient is

A 0.67

B 1.25

C 1.50

D 1.88.

15. Which of the following would **not** be suitable to act as a buffer solution?

A Boric acid and sodium borate

B Nitric acid and sodium nitrate

C Benzoic acid and sodium benzoate

D Propanoic acid and sodium propanoate

16. Which of the following 0.01 mol l^{-1} aqueous solutions has the highest pH value?

A Sodium fluoride

B Sodium benzoate

C Sodium propanoate

D Sodium methanoate

17. Which of the following graphs shows the variation in $\Delta G°$ with temperature for a reaction which is always feasible?

A
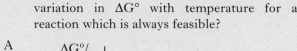

B
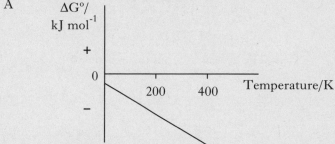

C
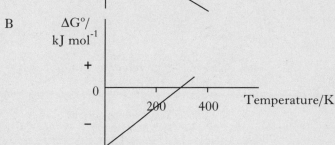

D
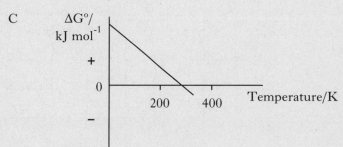

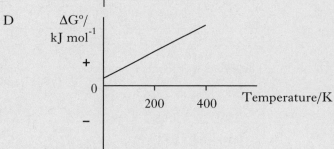

18. When water evaporates from a puddle which of the following applies?

   A ΔH positive and ΔS positive

   B ΔH positive and ΔS negative

   C ΔH negative and ΔS positive

   D ΔH negative and ΔS negative

19. For which of the following reactions would the value of $\Delta G^\circ - \Delta H^\circ$ be approximately zero?

   A $CaCO_3(s) \rightarrow CaO(s) + CO_2(g)$

   B $C(s) + H_2O(g) \rightarrow CO(g) + H_2(g)$

   C $Zn(s) + 2H^+(aq) \rightarrow Zn^{2+}(aq) + H_2(g)$

   D $Cu^{2+}(aq) + Mg(s) \rightarrow Mg^{2+}(aq) + Cu(s)$

20. For the reaction

$$2NO(g) + Cl_2(g) \rightarrow 2NOCl(g)$$

the rate equation is

$$\text{rate} = k[NO][Cl_2].$$

The overall order of this reaction is

   A   1

   B   2

   C   3

   D   5.

21. The following data refer to initial reaction rates obtained for the reaction

$$X + Y + Z \rightarrow \text{products}$$

Run	Relative concentrations			Relative initial rate
	[X]	[Y]	[Z]	
1	1·0	1·0	1·0	0·3
2	1·0	2·0	1·0	0·6
3	2·0	2·0	1·0	1·2
4	2·0	1·0	2·0	0·6

These data fit the rate equation

   A   Rate = k[X]

   B   Rate = k[X][Y]

   C   Rate = k[X][Y]^2

   D   Rate = k[X][Y][Z]

22. Which of the following is a propagation step in the chlorination of methane?

   A $Cl_2 \rightarrow Cl\bullet + Cl\bullet$

   B $CH_3{}^\bullet + Cl\bullet \rightarrow CH_3Cl$

   C $CH_3{}^\bullet + Cl_2 \rightarrow CH_3Cl + Cl\bullet$

   D $CH_4 + Cl\bullet \rightarrow CH_3Cl + H\bullet$

23. The hydrolysis of the halogenoalkane $(CH_3)_3CBr$ was found to take place by an S_N1 mechanism.

The rate-determining step involved the formation of

   A

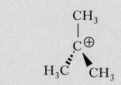

   B

   C

$$\left[\begin{array}{c} CH_3 \\ | \\ HO \text{----} C \text{----} Br \\ H_3C \quad CH_3 \end{array} \right]^{\oplus}$$

   D

$$\left[\begin{array}{c} CH_3 \\ | \\ HO \text{----} C \text{----} Br \\ H_3C \quad CH_3 \end{array} \right]^{\ominus}$$

24. $OH^- + CO_2 \rightarrow HCO_3^-$

$C_2H_4 + Br_2 \rightarrow C_2H_4Br^+ + Br^-$

Which substances act as electrophiles in the above reactions?

A OH^- and Br_2

B OH^- and C_2H_4

C CO_2 and Br_2

D CO_2 and C_2H_4

25.

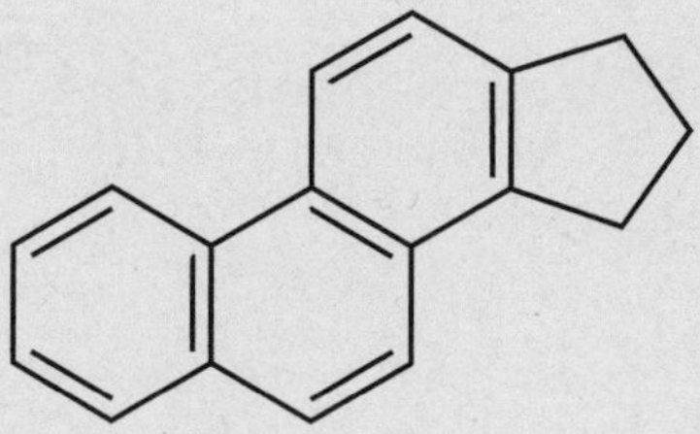

What is the molecular formula for the above structure?

A $C_{17}H_{11}$

B $C_{17}H_{14}$

C $C_{17}H_{17}$

D $C_{17}H_{20}$

26. Which line in the table is correct for the following hydrocarbon?

$$H_2C=C(CH_3)-C\equiv CH$$

	Number of σ bonds	Number of π bonds
A	4	3
B	8	5
C	10	2
D	10	3

27. When but-2-ene is shaken with an aqueous solution of chlorine in potassium iodide, the structural formula(e) of the product(s) is/are

A $CH_3-CH(I)-CH(I)-CH_3$

B $CH_3-CH(Cl)-CH(Cl)-CH_3$

C $CH_3-CH(Cl)-CH(I)-CH_3$ and $CH_2(I)-CH(Cl)-CH_2-CH_3$

D $CH_3-CH(Cl)-CH(I)-CH_3$ and $CH_3-CH(Cl)-CH(Cl)-CH_3$

28. Which of the following reacts with ethanol to form the ethoxide ion?

A $Na(s)$

B $Na_2O(s)$

C $NaCl(aq)$

D $NaOH(aq)$

29. Which of the following is **not** a correct statement about ethoxyethane?

A   It burns readily in air.

B   It is isomeric with butan-2-ol.

C   It has a higher boiling point than butan-2-ol.

D   It is a very good solvent for many organic compounds.

[Turn over

30. Which of the following esters gives a secondary alcohol when hydrolysed?

A

$(CH_3)_3C—O—\overset{\overset{\displaystyle O}{\|}}{C}—H$

B

$CH_3—O—\overset{\overset{\displaystyle O}{\|}}{C}—CH(CH_3)_2$

C

$(CH_3)_2CH—O—\overset{\overset{\displaystyle O}{\|}}{C}—CH_3$

D

$(CH_3)_2CHCH_2—O—\overset{\overset{\displaystyle O}{\|}}{C}—CH_3$

31. Which of the following compounds could **not** be oxidised by acidified potassium dichromate solution?

A CH_3CH_2CHO

B CH_3CH_2COOH

C $CH_3CH_2CH_2OH$

D $CH_3CH(OH)CH_3$

32. Which of the following will react with dilute sodium hydroxide solution?

A $CH_3CHOHCH_3$

B $CH_3CH=CH_2$

C CH_3COOCH_3

D $CH_3CH_2OCH_3$

33. Which of the following molecules is planar?

A Hexane

B Cyclohexane

C Chlorobenzene

D Methylbenzene (toluene)

34. Which of the following compounds is soluble in water and reacts with both dilute hydrochloric acid and sodium hydroxide solution?

A $C_2H_5NH_2$

B $C_6H_5NH_2$

C $C_2H_5NH_3Cl$

D $HOOCCH_2NH_2$

35. Which of the following reactions is least likely to take place?

A

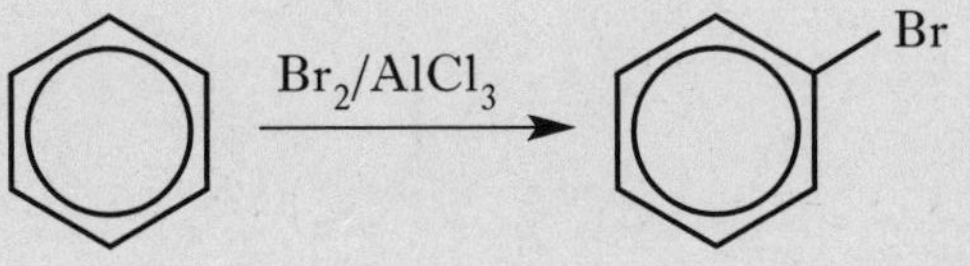

$\xrightarrow{Br_2/AlCl_3}$

B

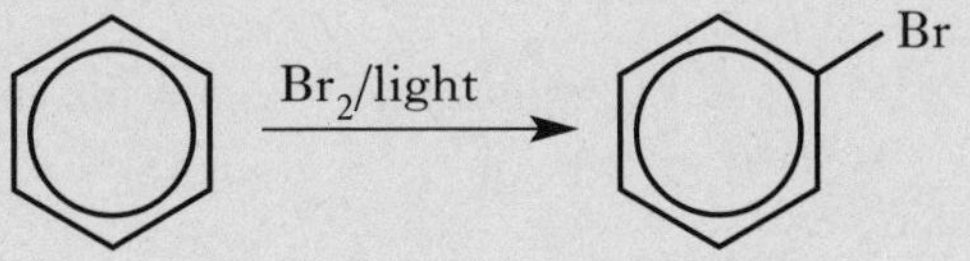

$\xrightarrow{Br_2/light}$

C

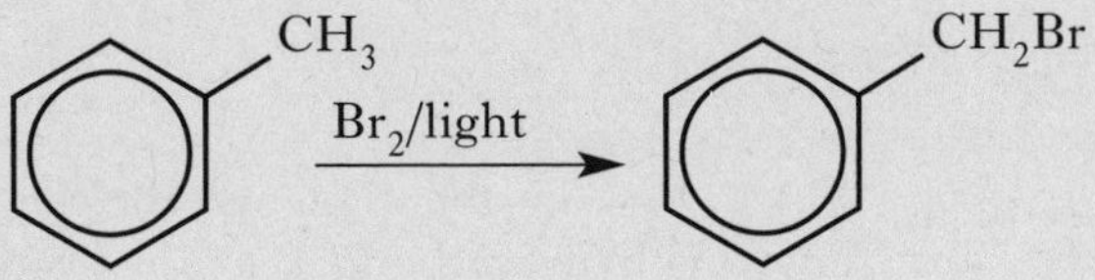

$\xrightarrow{Br_2/light}$

D

$\xrightarrow{H_2SO_4/SO_3}$

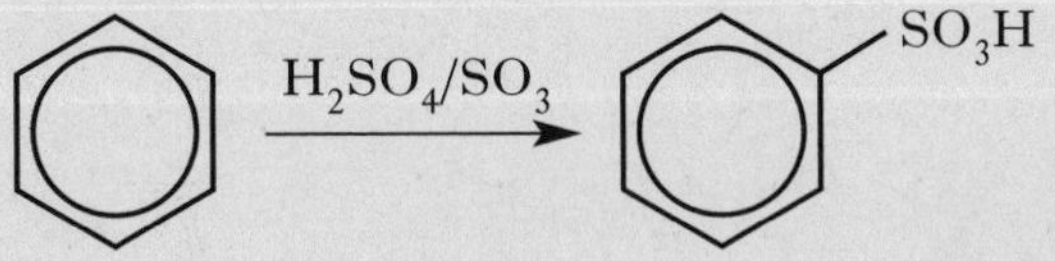

36. In which of the following pairs does an aqueous solution of the first compound have a higher pH than an aqueous solution of the second?

A

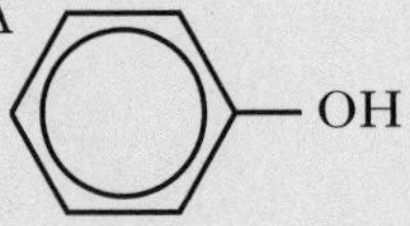

—OH　　and　　CH_3COOH

B

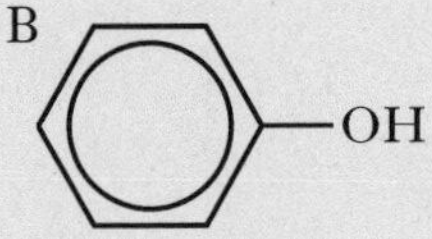

—OH　　and　　CH_3CH_2OH

C

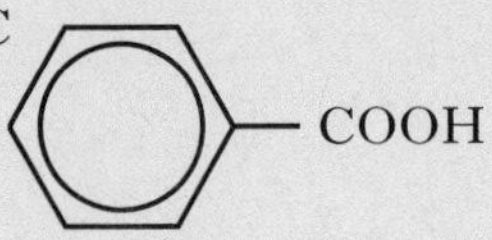

—COOH　　and　　$HOCH_2CH_2OH$

D

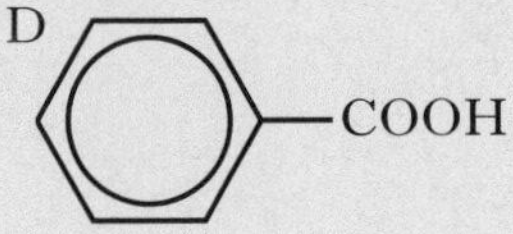

—COOH　　and　　CH_3OH

37. Which of the following bases is the strongest?

A　$C_2H_5NH_2$

B　$(C_2H_5)_2NH$

C　$C_6H_5NH_2$

D　$(C_6H_5)_2NH$

38. Which line in the table shows a pair of optical isomers?

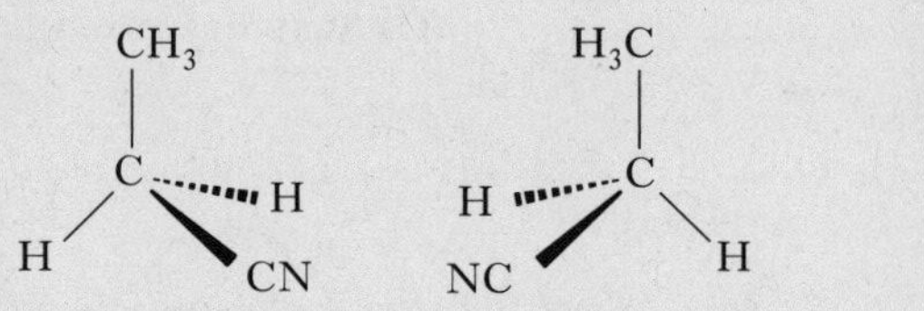

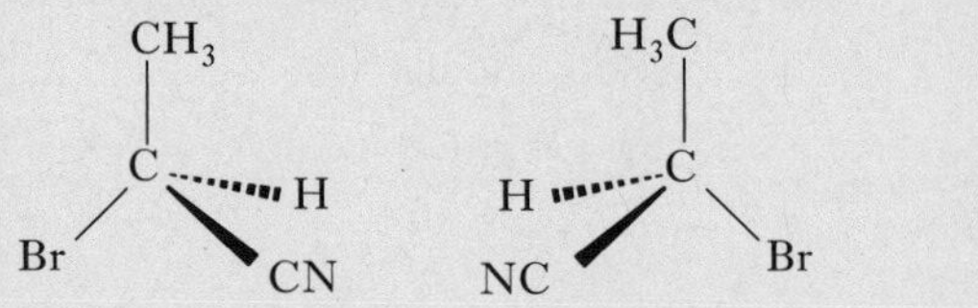

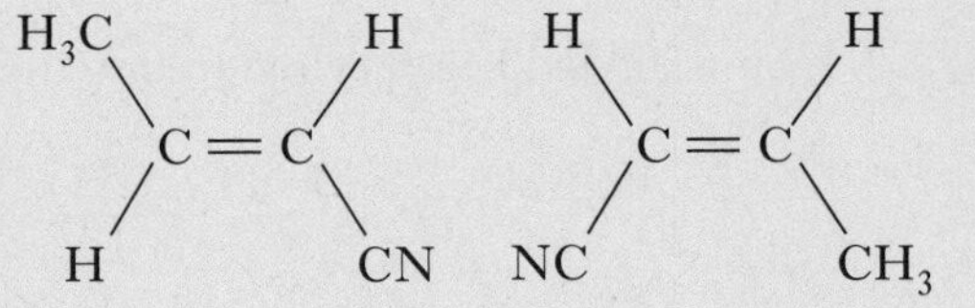

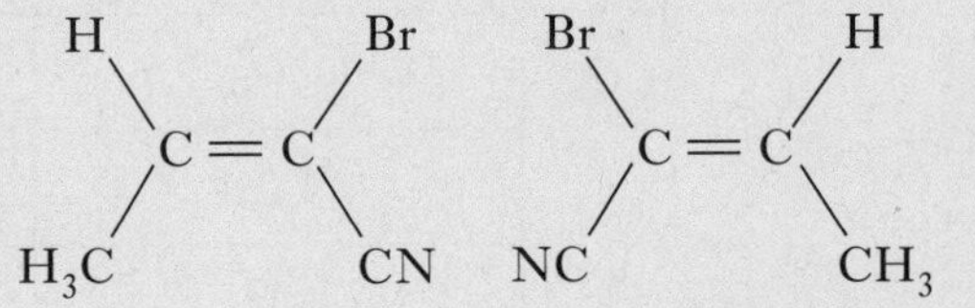

39.

HO—⟨ ⟩—I

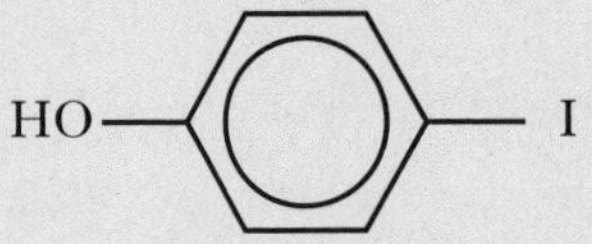

Which atom in the above structure would be located **most** readily using X-ray crystallography?

A　Carbon

B　Hydrogen

C　Iodine

D　Oxygen

40. Antihistamines act by inhibiting the action of the inflammatory agent histamine in the body.

Antihistamines can be described as

A　agonists

B　receptors

C　antagonists

D　pharmacophores.

[END OF SECTION A]

Candidates are reminded that the answer sheet for Section A MUST be placed INSIDE the front cover of your answer book.

SECTION B

60 marks are available in this section of the paper.

All answers must be written clearly and legibly in ink.

Marks

1. A detector in a Geiger counter contains argon which ionises when nuclear radiation passes through it.

   (a) Write the electronic configuration for argon in terms of s and p orbitals.　**1**

   (b) The first ionisation energy of argon is $1530\,kJ\,mol^{-1}$.

      (i) Calculate the wavelength of the radiation, in nm, corresponding to this energy.　**3**

      (ii) Write the equation for the first ionisation of argon.　**1**

                       (5)

2. Iron(III) oxide can be reduced to iron using hydrogen.

$$Fe_2O_3(s) + 3H_2(g) \rightarrow 2Fe(s) + 3H_2O(g)$$

Substance	$\Delta H_f^{\circ}/kJ\,mol^{-1}$	$S^{\circ}/J\,K^{-1}mol^{-1}$
$Fe_2O_3(s)$	−822	90
$H_2(g)$	0	131
$Fe(s)$	0	27
$H_2O(g)$	−242	189

For the reduction of iron(III) oxide with hydrogen, use the data in the table to calculate

   (a) the standard entropy change, ΔS°　**1**

   (b) the standard enthalpy change, ΔH°　**1**

   (c) the theoretical temperature above which the reaction becomes feasible.　**2**

                       (4)

Marks

3. The diagram, which is not drawn to scale, represents the processes involved in a thermochemical cycle.

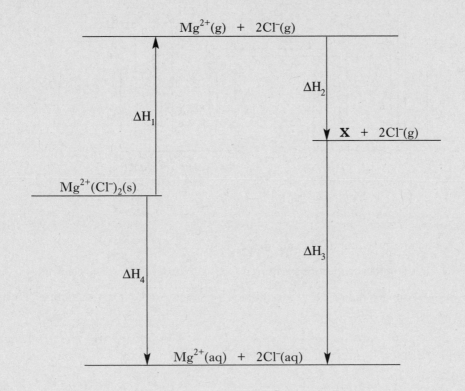

$Mg^{2+}(g) + 2Cl^-(g)$

ΔH_2

ΔH_1

$\mathbf{X} + 2Cl^-(g)$

$Mg^{2+}(Cl^-)_2(s)$

ΔH_3

ΔH_4

$Mg^{2+}(aq) + 2Cl^-(aq)$

(a) What should be written in place of **X** to complete the diagram?   1

(b) What name is given to the enthalpy change represented by ΔH_1?   1

(c) Calculate ΔH_3 using information from the Data Booklet.   1

(d) Calculate ΔH_4 using information from the Data Booklet.   1

                                                                                        (4)

4. (a) Using the mean bond enthalpy values given in the Data Booklet, calculate the enthalpy change, in $kJ\,mol^{-1}$, for the reaction

$$H_2(g) + \tfrac{1}{2}O_2(g) \rightarrow H_2O(g)$$ 3

    (b) The value given in the Data Booklet for the standard enthalpy of combustion of hydrogen is different to that calculated in part (a).

    Give the main reason for this difference.   1

                                                                                        (4)

[Turn over

Marks

5.

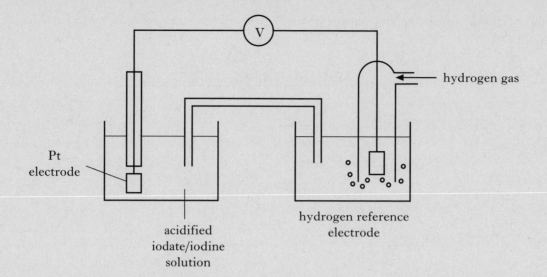

Pt electrode

acidified
iodate/iodine
solution

hydrogen gas

hydrogen reference
electrode

The above cell was set up under standard conditions.

(a) What are the three standard conditions required for the hydrogen reference electrode?   **1**

(b) Write an ion-electron equation for the reduction of iodate ions (IO_3^-) to iodine (I_2) in acidic conditions.   **1**

(c) If the E° value for the reduction of IO_3^- to I_2 is $1 \cdot 19\,V$, calculate the free energy change $\Delta G°$, in kJ per mole of IO_3^-, for the cell reaction.   **3**

(5)

6. When an ant bites, it injects methanoic acid (HCOOH).

(a) Methanoic acid is a weak acid.

$$HCOOH(aq) \; + \; H_2O(\ell) \; \rightleftharpoons \; HCOO^-(aq) \; + \; H_3O^+(aq)$$

(i) What is the conjugate base of methanoic acid?   **1**

(ii) Write the expression for the dissociation constant, K_a, of methanoic acid.   **1**

(b) (i) In a typical bite, an ant injects $3 \cdot 6 \times 10^{-3}\,g$ of methanoic acid.

Assuming that the methanoic acid dissolves in $1 \cdot 0\,cm^3$ of water in the body, calculate the concentration of the methanoic acid solution in $mol\,l^{-1}$.   **2**

(ii) Calculate the pH of this methanoic acid solution.   **2**

(6)

Marks

7. Iodine reacts with propanone as follows.

$$I_2 + CH_3COCH_3 \longrightarrow CH_3COCH_2I + HI$$

A possible mechanism for this reaction is

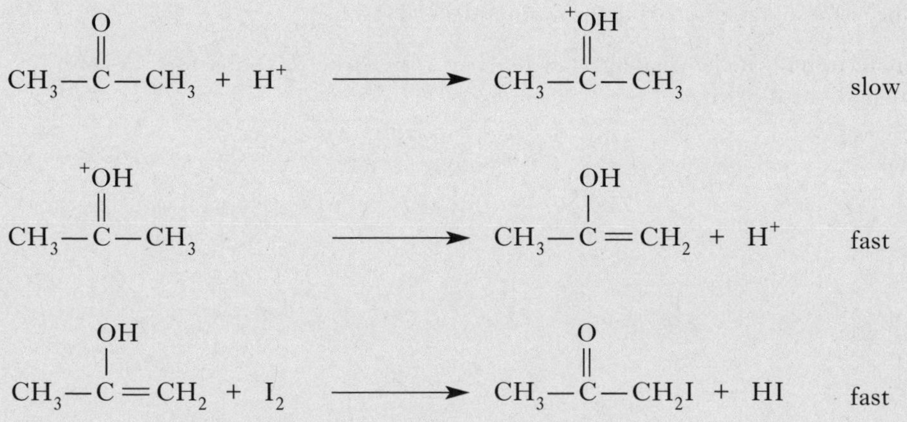

slow

fast

fast

(a) Write a rate equation for this reaction based on the above mechanism.   1

(b) What evidence indicates that the reaction is acid catalysed?   1

(c) In a PPA the reaction was followed by withdrawing samples at regular intervals and adding them to sodium hydrogencarbonate solution.

The concentration of iodine in these samples was then determined by titrating with a standard solution of sodium thiosulphate.

   (i) Why were the samples added to the sodium hydrogencarbonate solution?   1

   (ii) What indicator is used in the titration and what is the colour change at the end-point of the titration?   1

(4)

[Turn over

Marks

8. Nickel can be determined quantitatively in a number of ways.

   (a) The method used in a PPA is volumetric analysis in which a buffered solution of nickel(II) ions is titrated against a standard solution of a complexing agent.

   Which complexing agent is used?   **1**

   (b) Another way of determining nickel is by colorimetric analysis.

   Why would this be a suitable method of determining nickel(II) ions?   **1**

   (c) A third way of determining nickel depends on the fact that nickel(II) ions form a solid complex with butanedione dioxime.

butanedione dioxime   insoluble complex

   Using this method, a sample of a nickel(II) salt was accurately weighed and dissolved in water. To this solution, excess butanedione dioxime solution was added. The solid complex formed was filtered, washed and then heated in an oven to constant mass.

   (i) Butanedione dioxime can act as a ligand.

   What property of butanedione dioxime allows it to act as a ligand?   **1**

   (ii) What is the coordination number of the nickel(II) ion in the insoluble complex?   **1**

   (iii) Which type of quantitative analysis has been carried out using this method?   **1**

   (iv) During the process of heating to constant mass, the solid complex is cooled in a desiccator.

   Why is a desiccator used?   **1**

   (6)

Marks

9. Compound **W** reacts in two steps to form compound **Y**.

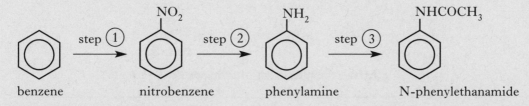

$$C_4H_9Br \xrightarrow{\text{①}} C_4H_{10}O \xrightarrow{\text{②}} C_4H_8O$$
$$\quad\mathbf{W}\qquad\qquad\qquad\mathbf{X}\qquad\qquad\qquad\mathbf{Y}$$

Y reacts with 2,4-dinitrophenylhydrazine solution (Brady's reagent) to form a yellow precipitate **Z**.

Y does not react with Fehling's solution, nor with Tollens' reagent.

(*a*) Identify compound **Y**.   1

(*b*) What type of reaction is occurring in step ①?   1

(*c*) What property of the yellow precipitate **Z** is measured and how is this used to confirm the identity of **Y**?   1

(*d*) Dehydration of compound **X** produces three unsaturated isomers of molecular formula C_4H_8. Two of these are **geometric** isomers.

Draw the structures of both **geometric** isomers and name each one.   2

                                                                                          (5)

10. N-Phenylethanamide can be prepared from benzene in three steps.

benzene $\xrightarrow{\text{step ①}}$ nitrobenzene $\xrightarrow{\text{step ②}}$ phenylamine $\xrightarrow{\text{step ③}}$ N-phenylethanamide

(with substituents NO_2, NH_2, $NHCOCH_3$ respectively)

(*a*) What chemicals are required to react with benzene to bring about step ①?   1

(*b*) What type of reaction occurs in step ②?   1

(*c*) Suggest a reagent which could be used to bring about step ③.   1

                                                                                          (3)

[Turn over

11. Spectra of an organic compound **A** are shown below.

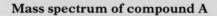

Mass spectrum of compound A

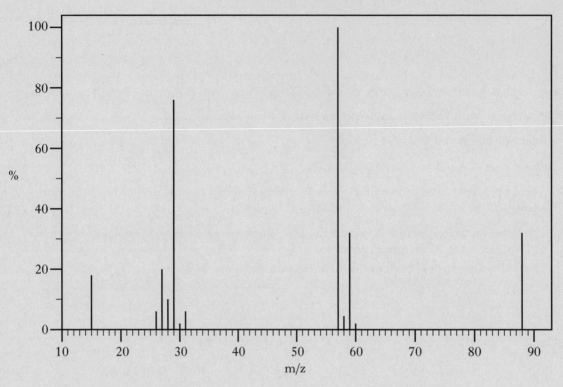

Infra-red spectrum of compound A

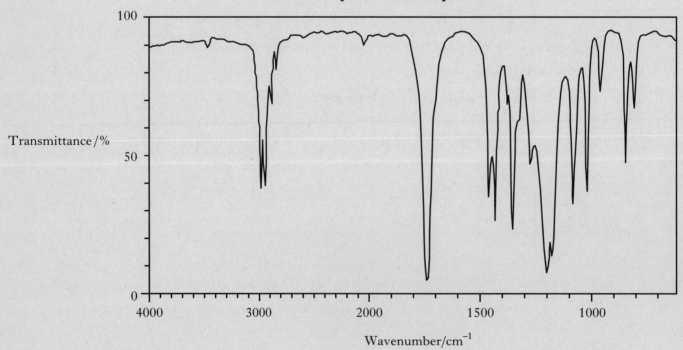

Marks

11. (continued)

Proton nmr spectrum of compound A

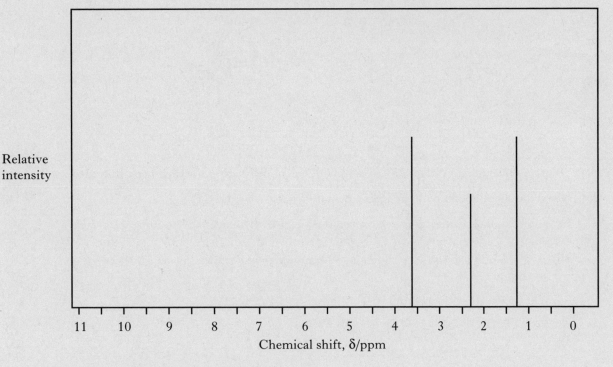

(a) Compound **A** has empirical formula C_2H_4O.

Using this information and the mass spectrum, deduce the molecular formula of **A**.   **1**

(b) The absorption peak at $1745\ cm^{-1}$ in the infra-red spectrum can be used to help identify **A**.

   (i) Which bond is responsible for this absorption?   **1**

   (ii) Which type of compound is **A**?   **1**

(c) Draw the structure of the ion fragment responsible for the peak at m/z 57 in the mass spectrum.   **1**

(d) Considering all the evidence, including the proton nmr spectrum, name compound **A**.   **1**

                                                        (5)

[Turn over

Marks

12. Many interhalogen compounds exist. Two of these are iodine pentafluoride and iodine heptafluoride.

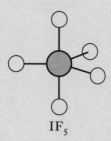

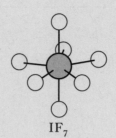

IF$_5$ IF$_7$

(a) What are the oxidation states of iodine in iodine pentafluoride and iodine heptafluoride?   **1**

(b) Name the shape adopted by the iodine pentafluoride molecule.   **1**

(c) In iodine heptafluoride, there are seven I–F bonds in which iodine uses sp^{3}d^3 hybrid orbitals.

Suggest which hybrid orbitals iodine uses in iodine pentafluoride, in which there are five I–F bonds.   **1**

(d) Another interhalogen compound, ClF$_5$, exists but ClF$_7$ does not.

Suggest a reason why ClF$_7$ does not exist.   **1**

(4)

Marks

13. A superconductor, **X**, with a critical temperature of 95 K, was prepared by heating yttrium oxide, barium carbonate and copper oxide at high temperatures.

(*a*) Copy the axes shown and sketch a graph to show how the **electrical resistance** of **X** varies with temperature.

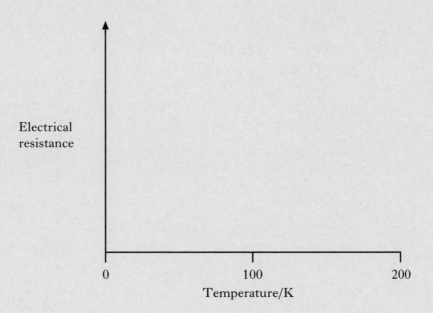

1

(*b*) (i) **X** contains 13·4% yttrium, 41·2% barium, 28·6% copper and 16·8% oxygen.

Assuming that the relative atomic mass of yttrium is 88·9, show by calculation that the empirical formula for **X** is $YBa_2Cu_3O_7$.

2

(ii) Assuming that the oxidation states of yttrium, barium and oxygen are +3, +2 and −2 respectively, calculate the **average** oxidation state of copper in **X**.

1

(iii) When all the copper(III) initially present in **X** is reduced to copper(II), compound **Z** is produced. The oxidation states of the other three elements do not change nor does the mole ratio of the **metals**.

Suggest an empirical formula for **Z**.

1

(5)

[END OF QUESTION PAPER]

[BLANK PAGE]